INORGANIC SYNTHESES

Volume XXI

Editor-in-Chief

JOHN P. FACKLER, JR.

Department of Chemistry
Case Western Reserve University

••

INORGANIC
SYNTHESES

Volume XXI

A Wiley-Interscience Publication
JOHN WILEY & SONS

New York   Chichester   Brisbane   Toronto   Singapore

QD151
C1
I5
vol. 21

Published by John Wiley & Sons, Inc.

Library of Congress Catalog Number: 39-23015

ISBN 0-471-86520-6

Printed in the United States of America

10 9 8 7 6 5 4 3 2 1

PREFACE

Inorganic Syntheses continues to fulfill a vital role for inorganic chemists and other users of inorganic and organometallic materials. Twenty volumes and 42 years have elapsed since Harold S. Booth edited the first volume in the series. While I did not have the opportunity to know Booth, his ghost occasionally returns to these halls at Case Western Reserve University to remind us of the great forward strides Inorganic Chemistry has made since World War II. Who would have known then that chromous acetate (*Inorganic Syntheses*, Vol. I, 1939, p. 122) contains a metal-metal quadruple bond or that sulfur hexafluoride (*Inorganic Syntheses*, Vol. I, 1939, p. 121) would find such an important use as an electrical insulator?

The objective of *Inorganic Syntheses* remains the same today as it was in 1939 when Roger Adams wrote: "The objective is to provide sufficient detail for each preparation so that a chemist of ordinary experience may duplicate the results on first trial." As Editor-in-Chief of Volume XXI, I have really learned what this means. Several submitted syntheses have, indeed, failed to meet this criterion, even though there was nothing suspect prior to actual checking.

The community of chemists who check these syntheses perform an invaluable service. I am truly grateful for their help and extend to them my personal thanks. I encourage you to let editors of future volumes know if you are willing to check various types of syntheses. Your help will be appreciated and the service you render will be most beneficial to future generations of chemists.

This volume has been divided (arbitrarily) into seven chapters. Chapter 1 contains some interesting syntheses of sulfur- and selenium-containing coordination compounds and clusters including the synthesis of the $[Fe_4S_4(SR)_4]^{2-}$ species which R. H. Holm so elegantly identified as models for active sites in nonheme iron proteins. Compounds formed from $S_2{}^{2-}$ and $S_5{}^{2-}$ are described. A convenient synthesis of CSe_2 is reported. Several additional compounds relevant to biological sulfur chemistry are described.

Dinuclear and polynuclear compounds continue to be synthesized in increasingly direct ways. Chapter 2 contains metal-metal bonded species such as those containing Pd—Pd or Mo≡Mo and W≡W bonds, mixed metal clusters, and compounds relevant to CO catalysis.

Organometallic species again occupy a major section of this volume, Chapter 3. A chapter on coordination compounds, Chapter 4, also appears. It includes the first checked description of CO_2 complexes of Ir. The tris(2,2'-bipyridine) ruthenium(II) chloride, well studied for its photochemical properties, is also reported. Chapter 5 is devoted to one-dimensional tetracyanoplatinate complexes. The tetrachlorooxotechretate(V) anion synthesis is described in Chapter 6 along with several other interesting stoichiometrically uncomplicated species. The use of technetium in medicine for imaging purposes has prompted renewed interest in this radioactive element. Several ligand syntheses are presented in Chapter 7, including the synthesis of pentamethylcyclopentodiene. A description of the synthesis of the useful one-electron reducing reagent VCl_2 is also presented.

In addition to thanking Fred Basolo and the Editorial Board who have generally encouraged me and given me help when needed, I wish to specifically thank Al Ginsberg, William Seidel, Steve Ittel, and Therold Moeller for the extra effort they have made to read or check manuscripts. Katherine Fackler helped with the initial organization of the volume and the indexing. Warren H. Powell and Thomas E. Sloan of the Chemical Abstracts Service again have kept the nomenclature up-to-date and have improved the English found in many of the manuscripts. I appreciate the help of Du Shriver, J. Worrell, M. Dolores Abood, and Anthony M. Mazany in proofreading.

Inorganic Syntheses first attracted my attention as an undergraduate at Vaparaiso University; but when I shared a laboratory as a graduate student with Professor Ralph C. Young at M.I.T., an early Associate Editor, I learned what *Inorganic Syntheses* was all about. My first personal copies of this work are due to his kindness.

Previous volumes of *Inorganic Syntheses* are available. Volumes I-XVI can be ordered from Robert E. Krieger Publishing Co., Inc., 645 New York Avenue (or P.O. Box 542), Huntington, NY 11743; Volume XVII is available from McGraw-Hill, Inc.; and Volumes XVIII-XX can be obtained from John Wiley & Sons, Inc.

JOHN P. FACKLER, JR.

Cleveland, Ohio
January 1982

NOTICE TO CONTRIBUTORS
AND CHECKERS

The *Inorganic Syntheses* series is published to provide all users of inorganic substances with detailed and foolproof procedures for the preparation of important and timely compounds. Thus the series is the concern fo the entire scientific community. The Editorial Board hopes that all chemists will share in the responsibility of producing *Inorganic Syntheses* by offering their advice and assistance in both the formulation of and the laboratory evaluation of outstanding syntheses. Help of this kind will be invaluable in achieving excellence and pertinence to current scientific interests.

There is no rigid definition of what constitutes a suitable synthesis. The major criterion by which syntheses are judged is the potential value to the scientific community. An ideal synthesis is one that presents a new or revised experimental procedure applicable to a variety of related compounds, at least one of which is critically important in current research. However, syntheses of individual compounds that are of interest or importance are also acceptable. Syntheses of compounds that are readily available commercially at reasonable prices are not acceptable.

The Editorial Board lists the following criteria of content for submitted manuscripts. Style should conform with that of previous volumes of *Inorganic Syntheses*. The introductory section should include a concise and critical summary of the available procedures for synthesis of the product in question. It should also include an estimate of the time required for the synthesis, an indication of the importance and utility of the product, and an admonition if any potential hazards are associated with the procedure. The Procedure should present detailed and unambiguous laboratory directions and be written so that it anticipates possible mistakes and misunderstandings on the part of the person who attempts to duplicate the procedure. Any unusual equipment or procedure should be clearly described. Line drawings should be included when they can be helpful. All safety measures should be stated clearly. Sources of unusual starting materials must be given, and, if possible, minimal standards of purity of reagents and solvents should be stated. The scale should be reasonable for normal laboratory operation, and any problems involved in scaling the procedure either up or down should be discussed. The criteria for judging the purity of the final product should be delineated clearly. The section on Properties should supply and discuss those physical and chemical characteristics that are relevant to judging the purity of the product and to permitting its handling and use in an intelligent manner. Under References, all pertinent literature citations should be listed in order. A style sheet is available from the Secretary of the Editorial Board.

The Editorial Board determines whether submitted syntheses meet the general specifications outlined above. Every synthesis must be satisfactorily reproduced in a laboratory other than the one from which it was submitted.

Each manuscript should be submitted in duplicate to the Secretary of the Editorial Board, Professor Jay H. Worrell, Department of Chemistry, University of South Florida, Tampa, FL 33620. The manuscript should be typewritten in English. Nomenclature should be consistent and should follow the recommendations presented in *Nomenclature of Inorganic Chemistry*, 2nd Ed., Butterworths & Co, London, 1970 and in *Pure Appl. Chem.*, 28, No. 1 (1971). Abbreviations should conform to those used in publications of the American Chemical Society, particularly *Inorganic Chemistry*.

Chemists willing to check syntheses should contact the editor of a future volume or make this information known to Professor Worrell.

TOXIC SUBSTANCES AND LABORATORY HAZARDS

The attention of the user is directed to the notices under this heading on page ix of Volume XIX and pages xv-xvii in Volume XVIII. It cannot be redundant to stress the ever-present need for the experimental chemist to evaluate procedures and anticipate and prepare for hazards. Obvious hazards associated with the preparations in this volume have been delineated, but it is impossible to forsee and discuss all possible sources of danger. Therefore the synthetic chemist should be familiar with general hazards associated with toxic, flammable, and explosive materials. In light of the primitive state of knowledge of the biological effects of chemicals, it is prudent that all the syntheses reported in this and other volumes of *Inorganic Syntheses* be conducted with rigorous care to avoid bodily contact with reactants, solvents, and products.

CONTENTS

Chapter One   METAL CHALCOGENIDE COMPOUNDS

Chapter Two DINUCLEAR AND POLYNUCLEAR COMPOUNDS

11. Binuclear Transition Metal Complexes Bridged by Methylene-
    bis(diphenylphosphine) 47
    A. Dichlorobis-μ-[methylenebis(diphenylphosphine)] -dipalladium(I)
       (Pd-Pd) 48
    B. μ-Carbonyl-dichlorobis[methylenebis(diphenylphosphine)] di-
       palladium(I). 49
    C. Tetrakis(1-isocyanobutane)bis[methylenebis(diphenylphos-
       phine)] dirhodium(I)bis[tetraphenylborate(1-)] 49
12. Synthesis of Dimethylamido Compounds Containing Metal-to-Metal
    Triple Bonds Between Molybdenum and Tungsten Atoms 51
    A. Hexakis(dimethylamido)dimolybdenum (Mo≡Mo), $Mo_2(NMe_2)_6$ 54
    B. Dichlorotetrakis(dimethylamido)dimolybdenum (Mo≡Mo),
       $Mo_2Cl_2(NMe_2)_4$ 56
13. Tetranuclear Mixed-Metal Clusters 57
    A. Tridecacarbonyldihydridoirontriruthenium, $FeRu_3H_2(CO)_{13}$. 58
    B. μ-Nitrido-bis(triphenylphosphorus)(1+)tridecacarbonylhydrido-
       irontriruthenate(1-), $[((C_6H_5)_3P)_2N]$ $[FeRu_3H(CO)_{13}]$ 60
    C. μ-Nitrido-bis(triphenylphosphorus)(1+)tridecacarbonylcobalt-
       triruthenate(1-), $[((C_6H_5)_3P)_2N]$ $[CoRu_3H(CO)_{13}]$ 61
    D. Tridecacarbonyldihydridoirontriosmium, $FeOs_3H_2(CO)_{13}$ 63
    E. Tridecacarbonyldihydridorutheniumtriosmium, $RuOs_3H_2(CO)_{13}$ 64
14. Bis[μ-nitrido-bis[triphenylphosphorus(1+)] Tridecacarbonyl-
    tetraferrate(2-) or [bis(triphenylphosphine)iminium
    tridecacarbonyltetraferrate] (2-)[PPN]$_2$[Fe$_4$(CO)$_{13}$] 66

Chapter Three ORGANOMETALLIC COMPOUNDS

15. Chlorobis(pentafluorophenyl)thallium(III) 71
    A. Preparation of Anhydrous Thallium(III) Chloride 72
    B. Preparation of (Pentafluorophenyl)lithium 72
    C. Preparation of Chlorobis(pentafluorophenyl)thallium(III) 72
16. (η^6-Hexamethylbenzene)ruthenium Complexes 74
    A. Di-μ-chloro-bis[chloro(η^6-1-isopropyl-4-methylbenzene)
       ruthenium(II)] 75
    B. Di-μ-chloro-bis[chloro(η^6-hexamethylbenzene)ruthenium(II)] . 75
    C. Bis(η^2-ethylene)(η^6-hexamethylbenzene)ruthenium(0) 76
    D. (η^4-1,3-Cyclohexadiene)(η^6-hexamethylbenzene)ruthenium(0) . 77

INORGANIC SYNTHESES

Volume **XXI**

Chapter One

METAL CHALCOGENIDE COMPOUNDS

1. SYNTHESIS OF (η^6-BENZENE)DICARBONYL(SELENOCAR-BONYL)CHROMIUM(0) AND PENTACARBONYL(SELENO-CARBONYL)CHROMIUM(0)

Submitted by IAN S. BUTLER,* ANN M. ENGLISH,* and KEITH R. PLOWMAN*
Checked by RUTH A. PICKERING† and ROBERT J. ANGELICI†

The activation of CX_2 (X = O, S, Se) by transition metals has received relatively little attention by comparison with the activation of CO_2 by nontransition elements, which has widespread application in organic synthesis. Recent reviews on CO_2[1] and CS_2[2-4] activation have revealed that these molecules can coordinate to transition metals to afford several different types of complexes. In one case, the CX_2 (X = S, Se) molecules act as sources of CX and lead to the formation of metal−CX bonds. This type of activation is of particular interest with respect to CSe_2 because carbon monoselenide, CSe, has never been isolated,[5] whereas CO and CS are well-characterized diatomic species. Studies on the activation of CSe_2 by transition metal complexes have recently been reported, and several stable metal selenocarbonyls can now be isolated, for example, (η^5-C_5H_5)M(CO)$_2$(CSe) (M = Mn,[6] Re[7]), cis-RuCl$_2$(CO)(CSe)[(C$_6$H$_5$)$_3$P]$_2$,[8] and Cr(CO)$_5$(CSe).[9] The

*Department of Chemistry, McGill University, 801 Sherbrooke St. West, Montreal, Quebec, Canada H3A 2K6.
†Department of Chemistry, Iowa State University, Ames, IA 50011.

and $Cr(CO)_5(CSe)$. Both these complexes are potentially extremely important because of the known catalytic activity of the analogous isoelectronic $(\eta^6\text{-}C_6H_6)Cr(CO)_3$ and $Cr(CO)_6$ species.[10,11] The synthesis of $(\eta^6\text{-}C_6H_6)Cr(CO)_2$-$(CSe)$ could readily be extended to the preparation of many other similar arene complexes.

■ **Caution.** *Carbon monoxide and metal carbonyls are highly toxic and must be handled in an efficient hood with care. The toxicity of CSe_2 is unknown. It has a vile smell and should always be handled with extreme care.*

A. $(\eta^6\text{-BENZENE})$DICARBONYL(SELENOCARBONYL)CHROMIUM(0)

$$(\eta^6\text{-}C_6H_6)Cr(CO)_3 + C_8H_{14} \xrightarrow{h\nu} (\eta^6\text{-}C_6H_6)Cr(CO)_2(C_8H_{14}) + CO$$

$$(\eta^6\text{-}C_6H_6)Cr(CO)_2(C_8H_{14}) + CSe_2 \longrightarrow (\eta^6\text{-}C_6H_6)Cr(CO)_2(CSe) + C_8H_{14} + Se$$

Procedure

Benzene (~ 250 mL) is distilled under nitrogen from sodium/benzophenone into a Pyrex ultraviolet irradiation vessel (capacity 350 mL) fitted with a water-cooled quartz finger containing a 100-W Hanovia high-pressure mercury lamp.* Following this, $(\eta^6\text{-}C_6H_6)Cr(CO)_3$ (1.0 g, 4.7 mmole)[12] and excess *cis*-cyclooctene (50 mL, 0.34 mole) are added to the benzene and the reaction mixture is well agitated with a stream of nitrogen. The reaction vessel is wrapped in aluminum foil and placed in an ice-water bath, and the ultraviolet lamp is turned on. (■ **Caution.** *Exposure of the eyes to ultraviolet light must be avoided at all times.*)

On irradiation, the color of the solution gradually turns dark wine-red owing to the formation of $(\eta^6\text{-}C_6H_6)Cr(CO)_2(C_8H_{14})$. The progress of the reaction is conveniently monitored by diluting a small sample of the reaction mixture with an equal volume of hexanes and by following changes in the CO stretching region (2150-1800 cm^{-1}) of the infrared spectrum of this mixture.† After about 2½ hours, no further changes are observed in the infrared spectrum, and the irradiation is terminated. The reaction mixture is allowed to stand for 30 minutes under a steady stream of nitrogen in order to remove any remaining CO. Carbon diselenide (1.0 g, 5.8 mmole)‡ is added to the reaction mixture and the

---

*The irradiation vessel has been described previously in *Inorganic Syntheses.*[6]

†In a solution infrared cell (pathlength 0.1 mm) fitted with NaCl windows. The bands observed are: $(\eta^6\text{-}C_6H_6)Cr(CO)_3$, $\sim$1970 and $\sim$1900 cm^{-1}; $(\eta^6\text{-}C_6H_6)Cr(CO)_2(C_8H_{14})$, $\sim$1900 and $\sim$1850 cm^{-1}.

‡CSe_2 is commercially available from Strem Chemicals Inc., Danvers, Mass., U.S.A. 01923 in 1.0 g ampules. A convenient synthesis is reported elsewhere in this volume.

ice-water bath is removed. The solution slowly turns dark yellow and some black decomposition product is evident. After standing for 2-3 hours, the reaction mixture is filtered under nitrogen through a medium-porosity, sintered glass filter to remove the dark-brown decomposition product. The benzene solvent and excess *cis*-cyclooctene are removed from the clear orange filtrate at room temperature on a rotary evaporator using a mechanical vacuum pump protected by a liquid nitrogen trap. The residue is washed repeatedly with small volumes of hexanes to remove any traces of *cis*-cyclooctene still adhering to the reaction product.

After drying completely, 0.6 g crude product is obtained [43% yield based on $(\eta^6\text{-}C_6H_6)Cr(CO)_3$]. Judging from the intensities of the $\nu(CO)$ bands of a sample of this crude product in CS_2, it is a 10/1 mixture of $(\eta^6\text{-}C_6H_6)Cr(CO)_2(CSe)$ and $(\eta^6\text{-}C_6H_6)Cr(CO)_3$. The selenocarbonyl complex is purified by thin-layer chromatography.* The crude complex is dissolved in a minimum quantity of diethyl ether and spotted onto six preparative thin-layer chromatography plates† (100 mg product per plate) and eluted with a 2/1 hexanes/diethyl ether solution. The trailing dark-yellow band of the selenocarbonyl product is quickly scraped off with a spatula. The scrapings are placed in a medium-porosity, sintered glass filter and washed with diethyl ether. The filtrate is reduced to dryness at room temperature on a rotary evaporator thus orange crystals are formed. An analytical sample of $(\eta^6\text{-}C_6H_6)Cr(CO)_2(CSe)$ (dec. 99°) is obtained by recrystallization from a hexanes/diethyl ether solution.

Anal. Calcd. for $C_9H_6O_2SeCr$: C, 39.0; H, 2.18; Se, 28.5; MW 278. Found: C, 39.2; H, 2.26; Se, 28.1; MW 278 (mass spectrum).

Properties

$(\eta^6\text{-}Benzene)$dicarbonyl(selenocarbonyl)chromium(0) is an air-stable, orange-yellow crystalline solid that darkens on continued exposure to light. It is soluble in CS_2, diethyl ether, acetone, chlorinated organic solvents, and benzene, but only slightly soluble in hydrocarbons. These solutions are stable for weeks if shielded from light. In the infrared spectrum in CS_2 solution, the bands due to CO and CSe stretching appear at 1975.6 (s) and 1932.0 (s) and at 1060.8 (s) cm⁻¹, respectively. Its ^{13}C nmr spectrum in CH_2Cl_2 solution at $-42°$ exhibits resonances at 100.9 (singlet, C_6H_6), 229.0 (singlet, $(CO)_2$), and 363.7 ppm (singlet, CSe) downfield positive relative to $(CH_3)_4Si$ ($\delta = 0.00$ ppm). The

---

*Since the selenocarbonyl product is photosensitive in solution, this procedure is performed with a minimum exposure to light in a darkened hood.

†A slurry of 80 g silica gel G (Macherey Nagel & Co., 516 Düren, West Germany) and water (170 ml) coats five 20 × 20 cm plates 1 mm thick. The plates are activated before use by heating them at 110°C for 1 hr. Precoated plates can be purchased from Canlab, 8655 Delmeade Rd., Town of Mount Royal, Montreal, Quebec, Canada H4T 1M3.

principal fragments in its mass spectrum at 70 eV are $(\eta^6\text{-}C_6H_6)Cr(CO)_2(CSe)^+$ (m/e 278), $(\eta^6\text{-}C_6H_6)Cr(CSe)^+$ (m/e 222), $Cr(CSe)^+$ (m/e 144), $(\eta^6\text{-}C_6H_6)Cr^+$ (m/e 130), and Cr^+ (m/e 52). There is no peak attributable to CSe^+ (m/e 92), consistent with the known instability[5] of CSe. In addition, there is no evidence for CSe loss except for the degradation of $[Cr(CSe)]^+$ to Cr^+, which suggests that the Cr—CSe bond is stronger than the Cr—CO bonds, in agreement with other physicochemical studies on metal selenocarbonyl complexes.[4]

B. PENTACARBONYL(SELENOCARBONYL)CHROMIUM(0)

$$(\eta^6\text{-}C_6H_6)Cr(CO)_2(CSe) + 3CO \longrightarrow Cr(CO)_5(CSe) + C_6H_6$$

Procedure

A solution of $(\eta^6\text{-}C_6H_6)Cr(CO)_2(CSe)$ [0.5 g, 1.6 mmole crude product from Section 1-A] in tetrahydrofuran (50 mL freshly distilled from sodium/benzophenone) is syringed into a stainless steel bomb (capacity 75 mL) fitted with a valve.* The valve is securely tightened to the bomb and, after degassing the solution by repeated freeze-thaw cycles at liquid nitrogen temperatures, CO gas (22 mmole) is introduced into the bomb from a calibrated vacuum manifold. The valve is closed and the bomb is placed in an oil bath and heated at 50° for 3-4 hours. (■ **Caution.** *The pressure in the bomb at room temperature, assuming that no CO gas dissolves, is approximately 20 atm.*) The bomb is again cooled to liquid nitrogen temperature before opening the valve,† and the CO gas remaining is slowly pumped off on a vacuum line in a well-ventilated area (hood!).

The bomb is then allowed to warm up to room temperature before the reaction mixture is syringed into a 250-mL round-bottomed flask. The flask is shielded from light as much as possible because the selenocarbonyl product is photosensitive. The tetrahydrofuran solvent is removed at -25 to $-22°$ (C_6H_5Cl/dry ice slush bath) on a rotary evaporator using a mechanical pump protected by a liquid nitrogen trap.‡ Sublimation of the residue under vacuum (25°/0.01 torr) onto an ice water-cooled finger inserted into the flask gives an analytically pure

---

*Stainless steel bomb (maximum operating pressure 1800 psig) and valve (¼ in. male × ¼ in. male) from Matheson of Canada Ltd., Whitby, Ontario, Canada LIN 5R9.

†Carbon monoxide has a vapor pressure of 400 torr at liquid nitrogen temperature ($-196°$).

‡The solvent must be removed at this temperature because of the high volatility of $Cr(CO)_5(CSe)$.

sample of $Cr(CO)_5(CSe)$ as deep-yellow crystals [0.23 g, 48% yield based on $(\eta^6\text{-}C_6H_6)Cr(CO)_2(CSe)$].*

Anal. Calcd. for C_6O_5SeCr: C, 25.5; Se, 27.9; MW 284. Found: C, 25.6; Se, 27.8; MW 284 (mass spectrum).

Properties

Pentacarbonyl(selenocarbonyl)chromium(0) is a highly volatile, deep-yellow crystalline solid that sublimes before melting. It is air stable but decomposes in a few hours on continued exposure to light, and is soluble without decomposition in all common organic solvents. In its infrared spectrum in *n*-hexane solution, the absorptions due to CO and CSe stretching modes are: $\nu(CO) = 2093$ (m) (a_1^{eq}), 2031 (m) (a_1^{ax}), and 2000 (vs) (e); $\nu(CSe) = 1077$ (s) (a_1) cm^{-1}, in agreement with the expected C_{4v} molecular symmetry. The ^{13}C nmr spectrum in CH_2Cl_2 solution at 27° shows resonances at 208.1 (singlet, *trans*-CO), 211.7 (singlet, *cis*-CO), and 360.7 ppm (singlet, CSe) downfield positive relative to $(CH_3)_4Si$ ($\delta = 0.00$ ppm). The resonances in its ^{17}O nmr spectrum in CH_2Cl_2 solution at 27° appear at 373.4 (singlet, *cis*-CO) and 385.3 ppm (singlet, *trans*-CO) downfield positive relative to $^{17}OH_2$ ($\delta = 0.00$ ppm). In both the ^{13}C and ^{17}O nmr spectra, the resonances for the *trans*-CO and *cis*-CO groups are in the 1:4 ratio expected. Its mass spectrum at 70 eV shows a similar fragmentation pattern to that described in Section 1-A above for $(\eta^6\text{-}C_6H_6)Cr(CO)_2(CSe)$: $[Cr(CO)_5(CSe)]^+$ (*m/e* 284) $\xrightarrow{-2CO}$ $[Cr(CO)_3(CSe)]^+$ (*m/e* 228) and stepwise loss of CO groups to give $[Cr(CSe)]^+$ (*m/e* 144). Again, there is no evidence for loss of CSe until $[Cr(CSe)]^+$ degrades to Cr^+ (*m/e* 52), nor is there a peak attributable to $[CSe]^+$. The chemistry of $Cr(CO)_5(CSe)$ is closely similar to that reported for $W(CO)_5(CS)$.[4]

References

1. N. S. Vyazankin, G. Razuvaev, and O. A. Kruglaya, in *Organometallic Reactions*, Vol. 5, E. I. Becker and M. Tsutsui (eds.), Wiley-Interscience, New York, 1975, p. 101.
2. I. S. Butler and A. E. Fenster, *J. Organometal. Chem.*, **66**, 161 (1974).
3. P. V. Yaneff, *Coord. Chem. Rev.*, **22**, 183 (1977).
4. I. S. Butler, *Acc. Chem. Res.*, **10**, 359 (1977).
5. R. Steudel, *Angew. Chem. Int. Ed.*, **6**, 635 (1967).

*Under more dilute reaction conditions (e.g., 0.2-0.3 g of the starting material), yields of 60-80% are obtained. It should also be emphasized that removal of traces of $(\eta^6\text{-}C_6H_6)Cr(CO)_3$ from the starting material is not necessary because $Cr(CO)_6$ is not formed under the mild reaction conditions used, and the low volatility of the tricarbonyl complex prevents it from subliming at 25°/0.01 torr.

6. I. S. Butler, D. Cozak, S. R. Stobart, and K. R. Plowman, *Inorg. Synth.*, **19**, 193 (1979).
7. I. S. Butler, D. Cozak, and S. R. Stobart, *Inorg. Chem.*, **16**, 1779 (1977).
8. G. R. Clark, K. R. Grundy, R. O. Harris, and S. M. James, *J. Organometal. Chem.*, **90**, C37 (1975).
9. A. M. English, K. R. Plowman, I. S. Butler, G. Jaouen, P. Lemaux, and J.-Y. Thépot, *J. Organometal. Chem.*, **132**, Cl (1977).
10. G. Jaouen and R. Dabard, in *Transition Metal Organometallics in Organic Synthesis*, Vol. 2, H. Alper, (ed.), Academic Press, New York, 1978.
11. M. S. Wrighton, D. S. Gimley, M. A. Schroeder, and D. L. Morse, *Pure Appl. Chem.*, **41**, 671 (1975).
12. M. D. Rausch, *J. Org. Chem.*, **39**, 1787 (1974).

2. DISELENOCARBAMATES FROM CARBON DISELENIDE

Submitted by WIE-HIN PAN* and JOHN P. FACKLER, JR.*
Checked by D. M. ANDERSON,† S. G. D. HENDERSON,† and T. A. STEPHENSON†

In 1972 a vastly improved synthesis was reported for the preparation[1,2] of CSe_2 by Henriksen and Kristiansen.[3] However, their procedure is designed for rather large amounts (400-500 g). Since CSe_2 readily polymerizes,[4] we felt it desirable to develop a procedure wherein smaller laboratory quantities of CSe_2 could be handled. We sought 1-10 g amounts of CSe_2 derivatives such as the dialkyldiselenocarbamates. Isolation of CSe_2 also seemed undesirable due to the foul smelling nature of the material and the toxicity of volatile species containing selenium.[5,6] The following syntheses are based on the procedures developed for the preparation of CSe_2 in a CH_2Cl_2 solution.

Mixed-Ligand Diselenocarbamates

The synthesis of the unsymmetrical Ni complexes of the type $Ni(S_2CNR_2)PR'_3X$, where R = Et or Me, R$'$ = *n*-Bu or Ph, and X = Cl, Br, or I, were first reported by Maxfield[7] in 1970. Subsequently, Fackler et al.[8] and Cornock and Stephenson[9] reported preparations of the Pd and Pt analogs. Complexes of this type provide an interesting series of compounds, which have been found to undergo substitution reactions quite readily[10,11] and hence are useful starting materials. The availability of the diseleno analogs allows comparisons to be made

*Department of Chemistry, Case Institute of Technology, Case Western Reserve University, Cleveland, OH 44106.
†Department of Chemistry, University of Edinburgh, Edinburgh EH9 3JJ, Scotland.

between the donor properties of Se and S in complexes of this type. In addition, the diseleno compounds are amenable to study by [77]Se nmr spectroscopy.[12]

Several approaches have been used to prepare mixed-ligand dithiocarbamates. For example, nickel complexes can be prepared[8] by mixing together $NiCl_2 \cdot 6H_2O$, trialkylphosphine, and sodium dialkyldithiocarbamate, all in stoichiometric amounts, in ethanol. The Pd complexes can be prepared[8] by adding the appropriate phosphine to the halide bridged dimer $[PdX(S_2CNR_2)]_2$. A general approach[9] is to react the sodium dialkyldithiocarbamate with a chloride-bridged dimer $[M(PR_3)Cl_2]_2$ (M = Ni, Pd, or Pt). Also, Tanaka and Sonoda[11] have prepared the selenothiocarbamato analog, $Pd(SSeCNR_2)PR_3Cl$, by reacting $[Pd(PR_3)Cl_2]_2$ with $(CH_3)_2SnCl(SSeCNR_2)$. An alternative approach has been reported[13] which has found general use in the syntheses of the diseleno compounds. This method, applicable also to the syntheses of the dithio analogs, is described here. The starting materials required are $M(PR_3)_2Cl_2$ and $M(Se_2CNR_2)_2$. The former are readily prepared according to literature methods.[14,15] The syntheses of the latter also have been previously described.[4,16,17]

A. CARBON DISELENIDE, CSe_2, SOLUTIONS

■ **Caution.** *CSe_2 has been classified as toxic.*[4]

$$2Se + CH_2Cl_2 \longrightarrow CSe_2 + 2HCl$$

Procedure

Since CSe_2 is toxic and malodorous,[4] it is essential that all work be conducted in a well-ventilated hood. Figure 1 is a diagram of the apparatus to be used.*

Grey Se powder (16-22 g) in a glass boat (14 × 1.5 cm) is placed in the Pyrex tube II, Fig. 1. Purified nitrogen gas (ca. 30 to 50 bubbles/sec) is passed through the system. Flask I is heated to 150-200°, section IIa of the tube to 375-400°, and section IIb of the tube to 575°. When sections IIa and IIb reach about 300 and 525° respectively, the dichloromethane is added dropwise into flask I at a rate of about 1 drop per second. Gradually, droplets of greenish-yellow liquid CSe_2 condense on the walls of the connector adapter III. A copious amount is formed as the temperature in IIa and IIb reaches 375 and 575°, respectively. As the reaction proceeds, a white mist is observed in the receiver flask IV. If the color of the mist appears reddish, indicating some unreacted Se vapor, the temperature of IIa is lowered. If the temperature in IIb is raised much above

---

*The checkers recommend that the end of tube II be a minimum of 10 mm in diameter. Further, they suggest that flask IV be connected to a reversed Dreschel bottle followed by a bottle containing 20 g KOH in 200 mL H_2O_2 (20% solution).

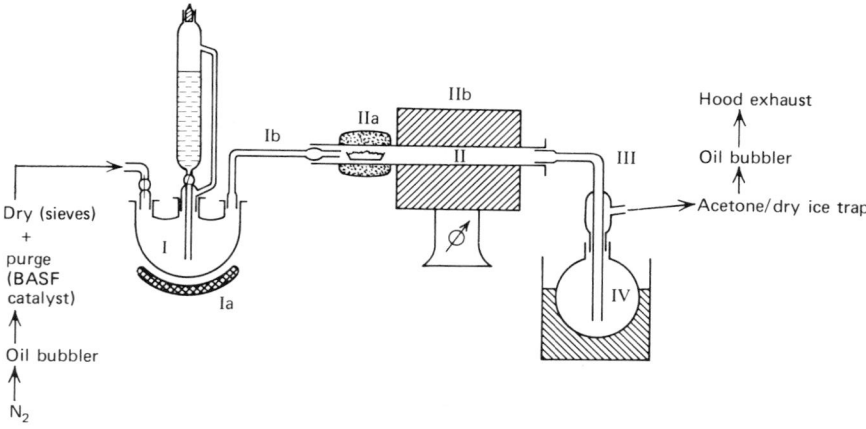

Fig. 1. Apparatus for the preparation of CSe_2. I. Dichloromethane vapor generator. 250-mL, three-neck flask with gas inlet, a funnel containing CH_2Cl_2 with a pressure-equalizing arm and a connecting tube (Ib). Ia is a 250-mL heating mantle. II. 60X3 cm Pyrex tube. Section IIa (ca. 15 cm long) is wound with nichrome wire and is insulated by packing with asbestos pulp. The tube is mounted on a Mutiple Unit Furnace (IIb). III. Vacuum connector adapter with one end extended as shown. IV. 250- or 500-mL round-bottom flask placed in a dry ice/acetone bath.

575°, decomposition of CSe_2 results and black carbon deposits form in the connector tube III and receiving flask IV. The reaction is complete in 2-3 hours with the cessation of the formation of the greenish-yellow droplets in III.

When the reaction is finished, heating of IIa and IIb is stopped. Flask IV is removed and stoppered. In its place is attached an empty flask; N_2 is passed continuously through the apparatus for another 2-3 hours. The contents of flask IV, usually contaminated with some black and red solids, are filtered through neutral alumina.* This is done in a well-ventilated hood or, even better, in a glove bag setup in a well-ventilated hood.

NOTE: It is advisable to have some saturated NaOH solution handy at all times to destroy any CSe_2 which may be spilled.

The clear, greenish-yellow liquid obtained upon filtration is a solution of CSe_2 in unreacted CH_2Cl_2. For the purpose of most syntheses, it is not necessary to remove the CH_2Cl_2. Rosenbaum et al.[16] have published a plot of the density

---

*The checkers recommend filtration using Schlenk tubes and catheter steel tubing to minimize odors.

(at 20°) of a mixture of CSe_2/CH_2Cl_2 versus the weight percent of CSe_2 in the mixture. Using this plot, a rough guide to the content of CSe_2 in the solution can be obtained. It generally indicates a yield of CSe_2 greater than 80% based on Se. In this form, the CSe_2 can be stored in a closed container at $-78°$ for 2-3 months with little, if any, decomposition.

After the tube has cooled, the apparatus is dismantled. The tube, with the boat in it, is filled with aqua regia and left standing overnight. This enables removal of the black deposits (presumably carbon) that have coated the glass. After removing the aqua regia, the tube is rinsed with water and vigorously scrubbed with a scouring pad and some cleanser.

Dialkyldiselenocarbamates

The various salts[16] of the dialkyldiselenocarbamates used to prepare metal diselenocarbamates[4,17] are formed using the CSe_2 in CH_2Cl_2. Literature methods[4,16,17] describing the reaction of CSe_2 with amines and base are closely followed.

Bis(phosphine)dihalo Metal Complexes

The preparations of $M(PR_3)_2Cl_2$, M = Ni, Pd, or Pt, and the related *cis*-$Pt(PPh_3)_2(CH_3)_2$ starting materials for the mixed-ligand diselenocarbamate syntheses follow literature methods.[14,15]

B. CHLORO(*N*,*N*-DIETHYLDISELENOCARBAMATO)(TRIETHYLPHOS-PHINE)NICKEL(II), NiCl(PEt$_3$)(Se$_2$CNEt$_2$)

$$Ni(Se_2CNEt_2)_2 + Ni(PEt_3)_2Cl_2 \rightarrow 2NiCl(PEt_3)(Se_2CNEt_2)$$

Bis(*N*,*N*-diethyldiselenocarbamato)nickel(II) (0.54 g; 1.0 mmole) is suspended in 75 mL acetone. Dichlorobis(triethylphosphine)nickel(II) (0.37 g; 1.0 mmole) is dissolved in 25 mL acetone. While stirring the $Ni(Se_2CNEt_2)_2$ suspension, the $Ni(PEt_3)_2Cl_2$ solution is added. The suspension slowly turns deep red. Stirring for about 30 minutes gives a dark-red solution. The volume is reduced by evaporation to 50 mL. Petroleum ether (100 mL) is added and the solution stored in the freezer for several hours. Very dark crystals are obtained which are filtered, washed with petroleum ether, and dried under vacuum. Yield - 0.75 g, 82%, mp, 92-5°. (At 80°, the color turns darker until at 90°, it is almost black. At 92°, the crystals melt giving a dark-red liquid.)[13]

C. CHLORO(N,N-DIETHYLDISELENOCARBAMATO)(TRIPHENYLPHOS-PHINE)PALLADIUM(II), PdCl(PPh$_3$)(Se$_2$CNEt$_2$)

$$Pd(Se_2CNEt_2)_2 + Pd(PPh_3)_2Cl_2 \rightarrow 2PdCl(Se_2CNEt_2)_2(PPh_3)$$

Bis(N,N-diethyldiselenocarbamato)palladium(II) (0.79 g, 1.33 mmole) and Pd(PPh$_3$)$_2$Cl$_2$ (0.94 g, 1.34 mmole) are mixed in 40 mL benzene (refluxed over sodium benzophenone before use). Under argon (or nitrogen), the suspension is stirred and heated at reflux. The suspension clears up to give an orange-red solution. Heating is stopped after 3 hours. The solution is filtered warm. Crystallization takes place upon cooling. Addition of petroleum ether (~10 mL) and cooling in the freezer yield more crystals. The crystals are filtered, washed with petroleum ether once, and dried under vacuum. The crystals are redissolved in the minimum amount of CHCl$_3$, and heptane is added to initiate crystallization. Yield 0.7 g (60%), mp 199-201°.[13]

D. CHLORO(N,N-DIETHYLDISELENOCARBAMATO)(TRIPHENYLPHOS-PHINE)PLATINUM(II), PtCl(PPh$_3$)(Se$_2$CNEt$_2$)

$$Pt(Se_2CNEt_2)_2 + Pt(PPh_3)_2Cl_2 \rightarrow 2PtCl(Se_2CNEt_2)(PPh_3)$$

Bis(N,N-diethyldiselenocarbamato)platinum(II) (0.58 g, 0.88 mmole) and Pt(PPh$_3$)$_2$Cl$_2$ (0.66 g, 0.83 mmole) are mixed in 35 mL benzene (refluxed over sodium benzophenone ketyl before use). The mixture is refluxed under nitrogen for about 7 hours. A clear, yellow solution is formed. The solution is evaporated to dryness in the hood. The yellow solid is recrystallized from CHCl$_3$/heptane. Yield, 1.0 g (81%), m.p. 205-8°.[13]

E. (N,N-DIETHYLDISELENOCARBAMATO)METHYL(TRIPHENYL-PHOSPHINE)PLATINUM(II), PtCH$_3$(PPh$_3$)(Se$_2$CNEt$_2$)

$$Pt(Se_2CNEt_2)_2 + Pt(PPh_3)_2(CH_3)_2 \rightarrow 2PtCH_3(Se_2CNEt_2)(PPh_3)$$

Bis(N,N-diethyldiselenocarbamato)platinum(II) (0.18 g, 0.27 mmole) and Pt(PPh$_3$)$_2$(CH$_3$)$_2$ (0.20 g, 0.27 mmole) are mixed in 30 mL benzene (refluxed over sodium benzophenone before use). Under argon, the mixture is stirred and heated at reflux. Heating is stopped after about 24 hours. The clear, yellow solution is reduced in volume to 10 mL. Petroleum ether is added until a light-yellow precipitate begins to form. The yellow product is recrystallized from $CH_3Cl/$ heptane. Yield 0.21 g (55%) mp 167-169°.[13]

Properties

The mixed-ligand diselenocarbamates of the nickel triad elements described here are planar, diamagnetic complexes. The nmr spectra[12,13] in solution are consistent with this geometry. The X-ray structures of $Pt(Se_2CNEt_2)(PPh_3)CH_3$,[18] $Pt(Se_2CNEt_2)(PPh_3)Cl$, and $Ni(Se_2CNEt_2)(PEt_3)Cl$[19] have been reported. Both solid-state structural work and solution nmr results, including [77]Se studies, demonstrate the structural trans influence,[18,19] $CH_3 \gg Cl$, wherein coupling constants trans to CH_3 are reduced and the Pt—Se distance lengthened compared with the analogous chloride complex.

References

1. H. G. Grimm and H. Metzer, *Chem. Ber.*, **69**, 1356 (1936).
2. D. J. G. Ives, R. W. Pittman, and W. Wardlow, *J. Chem. Soc.*, 1080 (1947).
3. L. Henriksen and E. S. S. Kristiansen, *Int. J. Sulfur Chem. A*, **2**, 133 (1972).
4. D. Barnard and D. T. Woodbridge, *J. Chem. Soc.*, 2922 (1961).
5. N. I. Sax, *Dangerous Properties of Industrial Materials*, 2nd ed. Reinhold Publishing Corp., New York, 1963, p. 575.
6. D. L. Klayman and W. H. H. Günther (eds.), *Organic Selenium Compounds: Their Chemistry and Biology*, Wiley-Interscience, New York, 1973.
7. P. L. Maxfield, *Inorg. Nucl. Chem. Lett.*, **6**, 693 (1970).
8. J. P. Fackler, Jr., I. J. B. Lin, and J. Andrews, *Inorg. Chem.*, **16**, 450 (1977).
9. M. C. Cornock and T. A. Stephenson, *J. Chem. Soc., Dalton*, 501 (1977).
10. J. A. McCleverty and N. J. Morrison, *J. Chem. Soc., Dalton*, 541 (1976).
11. N. Sonoda and T. Tanaka, *Inorg. Chim. Acta*, **12**, 261 (1975).
12. W.-H. Pan and John P. Fackler, Jr., *J. Am. Chem. Soc.*, **100**, 5783 (1979).
13. Wie-Hin Pan, John P. Fackler, Jr., and H.-W. Chen, *Inorg. Chem.* **20**, 856 (1981).
14. J. M. Jenkins and B. L. Shaw, *J. Chem. Soc., A*, 770 (1966).
15. C. R. Cousmaker, M. Hely-Hutchinson, J. R. Mellor, L. E. Sutton, and L. M. Venanzi, *J. Chem. Soc.*, 2705 (1961).
16. V. A. Rosenbaum, H. Kirchberg, and E. Leibnitz, *J. Prakt. Chem.*, **19**, 1 (1963).
17. K. A. Jensen and V. Krishnan, *Acta Chem. Scand.*, **24**, 1088 (1969).
18. H. W. Chen, J. P. Fackler, Jr., A. F. Masters, and Wie-Hin Pan, *Inorg. Chim. Acta*, **35**, L333 (1979).
19. W.-H. Pan, Ph.D. thesis, Case Western Reserve University, Cleveland, OH 1979.

3. POLYSULFIDE CHELATES

Submitted by RONALD A. KRAUSE,* ADRIENNE WICKENDEN KOZLOWSKI,[†] and JAMES L. CRONIN[‡]
Checked by ROSARIO DEL PILAR NEIRA[§]

At the turn of the century, Hoffman and Höchtlen[1,2] reported on crystalline polysulfides of heavy metals. However, their directions were very brief and no chemical characterization nor plausable structures were presented. Our interest in inorganic heterocyclic rings led us to investigate these and new polysulfide chelates.[3-5] Some of our detailed preparations are outlined here.

A. AMMONIUM POLYSULFIDE SOLUTION

$$2(NH_4)_2S + S_8 \rightarrow 2(NH_4)_2S_5$$

Procedure

1. In a fume hood, 100 mL $(NH_4)_2S$ solution (Mallinckrodt or J. T. Baker 21% solution) is stirred with 40 g sulfur. After ca. 1 hour, the excess sulfur is removed by filtration (10 g recovered) and the solution used as described below. The orange-red solution is stored in a closed container.

$$4NH_3(aq) + S_8 + 2H_2S \rightarrow 2(NH_4)_2S_5$$

2. A second method can be used for the generation of pentasulfide solutions, adapted from the method of Mills and Robinson.[6] Forty grams sulfur is mixed with 100 mL aqueous ammonia (conc.) in a fume hood. This solution is saturated with a stream of H_2S until all of the sulfur dissolves, and is then used as described below.

■ **Caution.** *Hydrogen sulfide is extremely poisonous. Use a good hood!*

B. AMMONIUM TRIS(PENTASULFIDO)PLATINATE(IV)

$$H_2PtCl_6 + 3(NH_4)_2S_5 + 2NH_3(aq) \rightarrow (NH_4)_2[Pt(S_5)_3] + 6NH_4Cl$$

*Department of Chemistry, The University of Connecticut, Storrs, CT 06268.
[†]Department of Chemistry, Central Connecticut State College, New Britain, CT 06050.
[‡]U.S. Borax and Chemical Corp., Los Angeles, CA 90010.
[§]Department of Chemistry, Case Western Reserve University, Cleveland, OH 44106.

Procedure

One gram $H_2PtCl_6 \cdot 6H_2O$(40% Pt; 2.05 mmole) in 10 mL of H_2O is added dropwise with continuous stirring to the solution obtained in Section 3-A. (Hood!) The solution immediately darkens; any initial precipitate of $(NH_4)_2PtCl_6$ rapidly redissolves. This solution is quickly filtered and stored in a closed container in a refrigerator. Solid obtained from this first filtration varies in quantity but is generally pure product.

After cooling the filtrate for 24 hours, the brick-red to maroon complex is isolated by filtration, washed two times with small portions of ice-cold water, and dried on the filter in vacuo over P_4O_{10}. The compound is then washed with CS_2 (hood!) to remove sulfur and redried as above.

This preparation gives a crystalline compound, but yields are variable depending on how much product is removed in the first filtration. Total yields average 75% (1.3 g); the second crop of solid can vary from 16 to 83% yield.

Anal. Nitrogen must be determined by the Dumas method (Kjeldahl giving very low results), sulfur by the Carius technique. Calcd. for $(NH_4)_2PtS_{15}$: N, 3.94; Pt, 27.40; S, 67.54. Found: N, 3.86; Pt, 27.22; S, 67.65.

Alternate Procedure for $(NH_4)_2[Pt(S_5)_3]$

To the solution obtained in Section 3-A2 is added 1.00 g H_2PtCl_6(2.05 mmole) (in 10 mL water), dropwise with stirring. In this method, no product is removed in the first filtration. The filtered solution is refrigerated in a closed container for 24 hours and the product isolated and treated as above. Yields in this procedure are more consistent, averaging 60%; however, this preparation yields a product which is invariably amorphous to X-rays.

C. TETRAPROPYLAMMONIUM BIS(PENTASULFIDO)PLATINATE(II)

$$(NH_4)_2Pt(S_5)_3 + 5KCN + 2[(C_3H_7)_4N]Br \rightarrow [(C_3H_7)_4N]_2[Pt(S_5)_2] +$$

$$5KSCN + 2NH_4Br$$

Procedure

One gram $(NH_4)_2Pt(S_5)_3$ (1.40 mmole) (from Section 3-B) is dissolved in 75 mL water at 70°, and 1.37 g solid KCN (21.0 mmole) (hood!) is added. The color of the solution immediately darkens; after 2 minutes, the solution is filtered, and 1.00 g $[(C_3H_7)_4N]Br$ (3.76 mmole) in 10 mL water is added. A red-orange precipitate forms at once; after cooling in ice, it is filtered, washed with water, and dried in vacuo over P_4O_{10}. Yield 1.02 g (82%).

Anal. (methods as above): Calcd. for $[(C_3H_7)_4N]_2Pt(S_5)_2$: C, 32.25; H, 6.31; N, 3.16; Pt, 21.98; S, 36.08. Found: C, 32.28; H, 6.71; N, 3.23; Pt, 21.05; S, 36.81.

Properties

The compound $(NH_4)_2Pt(S_5)_3$ is a brick-red crystalline solid soluble in water, acetone, and pyridine but insoluble in chloroform or diethyl ether. It shows no melting point but turns black at 145-150°.* The complex anion is readily precipitated from aqueous solution by large cations. The infrared spectrum (mineral oil mull) shows weak S—S modes at 490 and 450 cm^{-1}; additional weak bands are observed at 292 and 273 cm^{-1} in addition to the anticipated higher-energy NH_4^+ modes.

The compound $[(C_3H_7)_4N]_2Pt(S_5)_2$ is an orange-colored solid insoluble in water, slightly colors chloroform, and is soluble in acetone or acetonitrile. At 175-180°, the compound darkens; and it melts at 190-199°.* The infrared spectrum shows a number of bands due to $[(C_3H_7)_4N]^+$ below 1300 cm^{-1}; weak bands characteristic of the compound below 500 cm^{-1} are: 485, 450, 430, 330, 320, and 290 cm^{-1}.

Both compounds show poorly resolved spectra in the visible-UV region, with intense bands presumably of charge-transfer origin. Reaction with cyanide ion ultimately results in formation of the $Pt(CN)_4^{2-}$ ion in solution.

An X-ray crystal structure of $(NH_4)_2Pt(S_5)_3$ shows the platinum(IV) to be chelated in a pseudooctahedral fashion to three pentasulfide ligands.[7]

D. DISORDERED AMMONIUM BIS(HEXASULFIDO)PALLADATE(II)

$$PdCl_4^{2-} + 3S_5^{2-} + 2NH_4^+ \rightarrow (NH_4)_2PdS_{11} + S^{2-} + 4Cl^- + 3S$$

Procedure

An ammonium polysulfide solution is prepared as in Section 3-A above; to this solution is added 3.00 g K_2PdCl_4 (9.18 mmole) dissolved in 10 mL water. After shaking for 2 minutes, the initial black precipitate redissolves, and the solution is filtered through diatomaceous earth (filter aid or Celite) prewashed with ammonium hydroxide. The filtrate is stoppered and the reaction allowed to proceed 5 days at room temperature (ca. 22°). At this time, the solution is noticeably lighter in color and the product has precipitated. After filtration, the complex is washed with water, ethanol, diethyl ether, carbon disulfide, and

---

*Data provided by checker.

finally with diethyl ether before drying in vacuo over P_4O_{10}. Yield about 4.4 g (90%).

Anal. H, N, and S by commercial analysts as above. Palladium determined by treating the complex with aqueous bromine, heating to 100° for 5 hours, cooling and precipitating with dimethylglyoxime. Calcd. for $(NH_4)_2Pd\ S_{11}\cdot 2H_2O$: H, 2.27; N, 5.27; Pd, 20.03; S, 66.4. Found: H, 2.35; N, 5.16; Pd, 20.15; S, 66.55.

Properties

The complex is obtained as dark-red crystals insoluble in water. The structure[5] is different from that of other polysulfide complexes in that it consists of non-chelated S_6 chains linking planar palladium(II) ions. Chains appear to abort in their growth from the palladium, giving a defect structure accounting for the unusual formula.

This structure is most likely responsible for the fact that when the complex is dissolved in hot solvents (dimethylformamide, ethanol, pyridine, dimethylacetamide, or dimethyl sulfoxide), elemental sulfur and a dark residue are produced.

On heating, the solid does not melt but changes color from dark red to black at 165-227°.* The infrared spectrum below 500 cm^{-1} shows weak S—S modes at 480 and 440 cm^{-1} and additional bands at 360 (broad) and 300 cm^{-1}.

E. AMMONIUM TRIS(PENTASULFIDO)RHODATE(III)

$$RhCl_3 + 3NaCl \longrightarrow Na_3RhCl_6$$

$$Na_3RhCl_6 + 3(NH_4)_2S_5 \longrightarrow (NH_4)_3[Rh(S_5)_3] + 3NaCl + 3NH_4Cl$$

Procedure

Ammonium pentasulfide solution is prepared as described in Section 3-A. A solution containing 2.06 g $RhCl_3\cdot 3H_2O$ (7.82 mmole) and 1.50 g NaCl (25.7 mmole) in 15 mL water is added to the polysulfide solution causing an immediate darkening. After 1 minute of stirring, this mixture is filtered through a mat of diatomaceous earth prewashed with ammonium hydroxide. The filtrate is stoppered and allowed to stand (room temperature) for 2 days. Crystallization usually begins within 30 minutes, but maximum yield is obtained slowly. The product is isolated by filtration, washed with cold ammonium hydroxide, and dried in vacuo over P_4O_{10}. When dry, it is then washed with carbon disulfide. Yield about 3.0 g(60%).

---

*Data provided by checker.

Anal. H, N, S, and Rh by commercial analysts. Sulfur tends to be difficult (a modified Parr bomb procedure being most successful). Calcd. for $(NH_4)_3Rh(S_5)_3$: H, 1.90; N, 6.58; S, 75.4; Rh, 16.1. Found: H, 2.14; N, 6.23; S, 75.21; Rh, 16.2.

Properties

The compound is a dark red-purple dichroic cystalline material. In time, the crystals develop a thin coating of sulfur. The rhodium(III) complex, which appears to be structurally similar to the platinum(IV) compound, shows much greater reactivity in that sulfur is readily lost on dissolving in water, dimethylformamide, or pyridine. From water, a product may be precipitated which appears to contain the $[RhS_{10}]^-$ anion. Reaction with cyanide ultimately produces the thiocyanato complex.

On heating, the color of the solid changes from dark red-purple to black at 165-227°.* Below 500 cm^{-1}, the infrared spectrum shows weak S—S modes at 480 and 448 cm^{-1}; additional bands appear at 308 and 270 (broad) cm^{-1}.

*Data provided by checker.

References

1. K. A. Hoffman and F. Höchtlen, *Chem. Ber.*, 36, 3090 (1903).
2. K. A. Hoffman and F. Höchtlen, *Chem. Ber.*, 37, 245 (1904).
3. A. E. Wickenden and R. A. Krause, *Inorg. Chem.*, 8, 779 (1969).
4. R. A. Krause, *Inorg. Nucl. Chem. Lett.*, 7, 973 (1971).
5. P. S. Haradem, J. L. Cronin, R. A. Krause, and L. Katz, *Inorg. Chim. Acta*, 25, 173 (1977).
6. H. Mills and P. L. Robinson, *J. Chem. Soc.*, 2326 (1928).
7. P. E. Jones and L. Katz, *Chem. Commun.*, 842 (1967); *Acta Crystallogr.*, B25, 745 (1969).

4. DI-μ-CHLORO-TETRACHLORO-μ-(DIMETHYL SULFIDE)BIS(DIMETHYL SULFIDE)DINIOBIUM(III)

Submitted by MITSUKIMI TSUNODA† and LILIANE G. HUBERT-PFALZGRAF†
Checked by BRENT A. AUFDEMBRINK‡ and R. E. McCARLEY‡

The molecular chemistry of the lower oxidation states of niobium is still limited by its tendency to form polymeric insoluble clusters and by the lack of con-

†Laboratoire de Chimie Minérale Moléculaire, Université de Nice, Parc Valrose, 06034 Nice, France.
‡Department of Chemistry, Iowa State University, Ames, IA.

venient starting materials. The anhydrous trihalides are difficult to obtain, ill defined, nonstoichiometric, and too inert to be used for this purpose.[1] Until recently, very few molecular adducts of niobium(III) halides were known.[2] These were generally obtained from Nb_2Cl_{10} either through a two-step reductive process or by reduction of an isolated niobium(IV) intermediate. But these methods generally resulted in poor yields; $Nb_2Cl_6(SMe_2)_3$, for instance, was obtained in 28% yield from Nb_2Cl_{10}, by successive reactions, first with the borane-dimethyl sulfide adduct and then with an excess of Na/K alloy.[3] The recent utilization of magnesium turnings as reductant now provides a high-yield and convenient synthetic route to $Nb_2Cl_6(SMe_2)_3$.[4] The synthesis described below can be summarized by the following equation:

$$Nb_2Cl_{10} + 3Me_2S \xrightarrow[\text{CH}_2\text{Cl}_2\text{-Et}_2\text{O}]{\text{Mg excess}} Nb_2Cl_6(SMe_2)_3 + 2MgCl_2$$

Procedure

All manipulations are carried out under a nitrogen or argon oxygen-free atmosphere using standard septum, syringe, and Schlenk tube techniques.[5] Use of greaseless, Teflon-coated joints is recommended.

■ **Caution.** *The transfer of solvents and the reaction procedures are best conducted in an efficient fume hood.*

Niobium pentachloride, Nb_2Cl_{10} (4.58 g, 8.47 mmole), magnesium turnings (1.36 g, 55.9 mmole), and a magnetic stirring bar are placed in an argon-purged 500-mL round-bottomed flask. Dichloromethane, 100 mL, is added all at once. A solution of dimethyl sulfide (4.2 mL, 57.3 mmole) in 25 mL CH_2Cl_2 is added dropwise within 10 minutes to the stirred suspension, at room temperature. The solid dissolves rapidly, giving a dark red-brown solution. A solution of freshly distilled (from sodium benzophenone ketyl) diethyl ether (22.6 mL, 217.6 mmole)* in 25 mL CH_2Cl_2 is then added dropwise (8 minutes). After about 30 minutes of stirring, the reaction mixture becomes cloudy and the formation of $MgCl_2$ starts. The reaction medium turns dark violet after 24 hours of stirring at room temperature, and no change in the coloration is noted after 60 hours. The precipitate is then removed by filtration, using a small porosity frit, and washed with CH_2Cl_2 until the washings are colorless (about 60 mL in 4 portions). The filtrate and the washings are evaporated to dryness. The dark-violet residue is stirred with 80 mL CH_2Cl_2 for about 30 minutes and then filtered, leaving a violet-brown solid which is washed with about 60 mL CH_2Cl_2. The filtrate and the washing solutions are evaporated to dryness under vacuum. Removal of the last traces of solvent under high vacuum ($\approx$ 0.005 torr for 6

---

*The amount of diethyl ether is crucial and should correspond to at least 8 equivalents per niobium.

hours) at room temperature affords 4.40 g (88.1%)* of dark-violet crystals. The compound may be recrystallized from toluene-pentane.

Anal. Calcd. for $C_6H_{18}S_3Cl_6Nb_2$: C, 12.30; H, 3.10; Cl, 36.42; S, 16.41. Found: C, 12.15; H, 3.07; Cl, 36.71; S, 15.60.

Properties

Hexachlorotris(dimethyl sulfide)diniobium(III) forms dark-violet crystals that do not melt below 230°. The compound is stable in air for short periods of time, but can be stored for months under nitrogen or argon at 0°. It is insoluble in pentane, slightly soluble in diethyl ether, and more soluble in benzene, toluene, chloroform, dichloromethane, and (with rapid decomposition) acetonitrile.[4] Characterization of the product, which is diamagnetic, is most easily achieved by nmr spectroscopy. The ^1H nmr spectrum ($CDCl_3$, 25°) shows resonances at δ 2.63 (2, S—CH_3 terminal) and 3.34 (1, S—CH_3 bridging) ppm. The IR spectrum (Nujol) gives the following major absorptions below 1400 cm^{-1}: 1328 (m), 1310 (m), 1255 (m), 1090 (m), 1030 (s), 1020 (s), 980 (s), 975 (s), 800 (m), 685 (m), 355 (sh), 345 (vs), 322 (s), 318 (s), 285 (w), 238 (m), and 220 (m).

Ligand exchange reactions[3] give products such as $Nb_2Cl_6(dppe)_2$ [dppe = 1,2-ethanediylbis(diphenylphosphine)], $Nb_2Cl_6(diars)_2$ [diars = 1,2-phenylene-bis(dimethylarsine)], and $Nb_2Cl_6(triars)_2$ [triars = {2-[(dimethylarsino)methyl] -2-methyl-1,3-propanediyl}bis(dimethylarsine)]. $Nb_2Cl_6(SMe_2)_3$ is also a useful starting material for substitution reactions.[6]

References

1. D. L. Kepert, *The Early Transition Metals*, Academic Press, London, 1972.
2. E. T. Maas and R. E. McCarley, *Inorg. Chem.*, **12**, 1096 (1973).
3. A. D. Allen and S. Naito, *Can. J. Chem.*, **54**, 2948 (1976).
4. L. G. Hubert-Pfalzgraf and J. G. Riess, *Inorg. Chim. Acta*, **29**, L251 (1978), L. G. Hubert-Pfalzgraf, M. Tsunoda, and J. G. Riess, *Inorg. Chim. Acta*, **41**, 283 (1980).
5. D. F. Shriver, *The Manipulation of Air-Sensitive Compounds*, McGraw-Hill, New York, 1969.
6. L. G. Hubert-Pfalzgraf, M. Tsunoda, and D. Katoch, *Inorg. Chim. Acta*, **51**, 81 (1981).

*The checkers scaled down the synthesis to one-fourth the suggested amount and obtained a yield of ~35%.

5. (2-AMINOETHANETHIOLATO-*N*,*S*)BIS(1,2-ETHANE-DIAMINE)COBALT(III), (2-MERCAPTOACETATO-*O*,*S*)BIS(1,2-ETHANEDIAMINE)COBALT(III), AND RELATED COMPLEXES

Submitted by DENNIS L. NOSCO* and EDWARD DEUTSCH*
Checked by D. S. DUDIS†

Bis(1,2-ethanediamine)cobalt(III) complexes containing a chelated thiolato ligand have been synthesized by two routes. The first route, in which the chelating thiolato ligand displaces chloride from *cis*-[(en)$_2$CoCl$_2$]$^+$, results in poor yields.[1] Much superior yields are obtained by the second route in which chelating disulfides are reduced by (1,2-ethanediamine)cobalt(II) mixtures.[2] This second route is also advantageous in that the initial product is sufficiently pure for most subsequent uses, the total reaction time is less than 2 hours, the starting materials are inexpensive and readily available, and no intermediate cobalt(III) complex must be synthesized or isolated. The two bis(1,2-ethanediamine)thiolatocobalt(III) complexes detailed here are the cysteamine-*N*,*S* and mercaptoacetato-*O*,*S* complexes. However, this synthetic procedure can be readily extended to other chelating thiols such as thioisobutyric acid,[2d] thiolactic acid,[2d] 2-aminobenzenethiol,[2e] cysteine-*N*,*S*,[2f] and even selenium-containing compounds such as selenocysteamine[2g] and selenylacetic acid.[2g] The syntheses detailed herein have been revised and improved from those previously reported.[2a,2b]

A. (2-AMINOETHANETHIOLATO-*N*,*S*)BIS(1,2-ETHANEDIAMINE)COBALT(III) SALTS

$$2\,CoX_2 + (SCH_2CH_2NH_2\cdot HCl)_2 + 5en \xrightarrow[N_2]{H_2O} 2\,[(Co(en)_2(SCH_2CH_2NH_2)]X_2$$

$$+\ (enH_2)Cl_2$$

$$X = ClO_4^-,\, Cl^-,\, NO_3^-,\, I^-,\, \tfrac{1}{2}(SO_4^{-2})$$

$$en = H_2NCH_2CH_2NH_2$$

*Department of Chemistry, University of Cincinnati, Cincinnati, OH 45221.
†Department of Chemistry, Case Western Reserve University, Cleveland, OH 44106.

Procedure[2c]

- **Caution**. *Perchloric acid and perchlorate salts are strong oxidants and may explode or ignite on contact with organic material!*[3]

A 1-L Erlenmeyer flask containing 100 g (0.273 mole, 0.273 equiv) $Co(ClO_4)_2 \cdot 6H_2O$ dissolved in 375 mL distilled H_2O and a 500-mL Erlenmeyer flask containing 31.8 g (0.141 mole, 0.282 equiv) cystamine dihydrochloride $(SCH_2CH_2NH_2 \cdot HCl)_2$ dissolved in 200 mL distilled H_2O are simultaneously deaerated by vigorous bubbling of N_2 through the respective solutions for 30 minutes. A total of 4.56 g (0.760 mole) of neat 1,2-ethanediamine is then added slowly to the solution of cobalt(II) perchlorate to yield an orange precipitate. Deaeration of both solutions is continued for 10 additional minutes. At this point, the solution containing the disulfide is added in one portion to the cobalt(II)/1,2-ethanediamine mixture under anaerobic conditions. The orange precipitate rapidly dissolves, and the solution color changes from orange to brown. Vigorous N_2 bubbling is continued for 30 additional minutes.

At this point, the deaeration is stopped, and 50 mL concentrated (72%) perchloric acid ($HClO_4$) is added slowly to the brown solution.[3] The resulting solution is immediately filtered through a medium porosity frit. An additional 80 mL concentrated $HClO_4$ is added to the filtrate and the solution is refrigerated for 24 hours. The next day, black crystals are collected and washed with cold 95% ethanol. The crystals are dried over P_4O_{10} in a vacuum desiccator to yield 106 g (87%) yield of the title compound. The $[Co(en)_2(SCH_2CH_2-NH_2)](ClO_4)_2$ is recrystallized from 50-60° distilled water by cooling and/or addition of concentrated $HClO_4$ with cooling. Other salts are synthesized by starting with different cobalt(II) salts, such as $CoCl_2$, CoI_2, $CoSO_4$, and $Co(NO_3)_2$, and substituting the addition of saturated aqueous solutions of NH_4Cl, NaI (prepared immediately before use), $(NH_4)_2SO_4$, and NH_4NO_3, respectively, for addition of concentrated $HClO_4$.

Anal. Calcd. for $[CoC_6H_{22} N_5S](ClO_4)_2$: C, 15.89; H, 4.88; N, 15.42. Found: C, 15.94; H, 4.88; N, 15.59.

The single-crystal X-ray structure of the thiocyanate salt has been determined.[4]

Properties

Salts of the brown-black $[Co(en)_2(SCH_2CH_2NH_2)]^{2+}$ are stable indefinitely in air and are stable for days in solutions of dilute aqueous acid at moderate (60°) temperatures. The solid perchlorate salt will ignite in a Bunsen burner flame or on contact with a hot surface, but presents no hazard in solution. The perchlorate salt is the most soluble of those listed above, being very soluble in H_2O (0.4 *M* at 25°),[5] *N,N*-dimethylformamide (DMF), and dimethyl sulfoxide, soluble in triethyl phosphate, methanol, and tetrahydrothiophene 1,1-dioxide and slightly soluble in ethanol and acetone. The perchlorate salt can be dissolved in tetrahydrofuran (THF) if it is first dissolved in warm (50-60°) methane-sulfonic acid (CH_3SO_3H) and then added slowly to the THF. The visible/UV spectrum of this complex is characteristic[3c] with a shoulder at 600 nm (ϵ = 44 M^{-1} cm^{-1}), a *d-d* transition at 482 nm (ϵ = 140 M^{-1} cm^{-1}), and an intense sulfur-to-metal charge transfer band at 282 nm (ϵ = 14,000 M^{-1} cm^{-1}). The purity of the compound can be ascertained by determining the effective extinction coefficient of the 282-nm band. The visible-UV spectrum in solution is not a function of the counterion. This thiolato complex is used as a starting material for preparation of complexes in which the coordinated sulfur is modified to form coordinated S-bonded sulfenic acids,[2d] S-bonded thioethers,[6] S-bonded disulfides,[7] and S-bonded sulfenyl iodides.[8] It can also be used to standardize CH_3Hg^+ solutions.[9]

B. BIS(1,2-ETHANEDIAMINE)(2-MERCAPTOACETATO-*O,S*)COBALT(III) SALTS

$$2Co(ClO_4)_2 + 4en + (HOOCCH_2S)_2 \xrightarrow[N_2]{H_2O} 2[Co(en)_2(SCH_2COO)](ClO_4)$$

Procedure

Three flasks, a 500-mL Erlenmeyer flask containing 61.5 g (0.17 mole, 0.17 equiv) $Co(ClO_4)_2 \cdot 6H_2O$ in 100 mL distilled water, a 500-mL Erlenmeyer flask containing 21.8 g (0.363 mole) 1,2-ethanediamine in 196 mL distilled water, and a 250-mL Erlenmeyer flask containing 15.0 g (0.082 mole, 0.167 equiv) 2,2′-dithiobis(acetic acid) (obtained from Evans Chemetics, Inc.) $(HOOCCH_2S)_2$* in 90 mL distilled water, are all deaerated by bubbling N_2 vigorously through the solutions for 30 minutes. Deaeration of the Co(II) solution is continued as first the 1,2-ethanediamine solution, and then the 2,2′-dithiobis(acetic acid) solution are added to it. An orange color is caused by addition of the 1,2-ethanediamine; this changes to a dark red-brown color when the 2,2′-diethiobis(acetic acid) is added.

Deaeration is continued for 1 additional hour, and then the solution is allowed to stand for 7 hours at 25° and finally is refrigerated overnight at 4°C. The solution is then filtered through a medium glass frit, and the red-brown solid† is collected and washed with 50% ethanol-water. This solid is recrystallized from a minimum amount of warm (60°) 0.01 *M* $HClO_4$ by refrigeration. The resulting

*The 2,2′-dithiobis(acetic acid), commonly named dithioglycolic acid, used by the checker was synthesized from $HSCH_2COOH$ and SO_2Cl_2 in diethyl ether.
†The checker obtained purple crystals at this stage.

purple crystals are separated by filtration on a medium-porosity glass frit, washed with a small amount of 50% ethanol-water and then cold 95% ethanol, and dried in a vacuum desiccator to give a final yield of 18.0 g (30%). The perchlorate salt may be converted to the chloride salt by loading an aqueous solution (0.01 M H^+) of the perchlorate salt onto a Sephadex SP C-25 cation exchange column in the sodium form (Pharmacia), eluting the complex with 0.25 M NH_4Cl (0.01 M HCl), concentrating the eluent under vacuum (25°), and then refrigerating the concentrated eluent.

Anal. Calcd. for $[Co(en)_2(SCH_2COO)]$ $ClO_4 \cdot H_2O$: C, 22.33; H, 6.25; N, 17.36; S, 9.94; Cl, 10.99. Found: C, 22.19; H, 6.03; N, 17.30; S, 9.84; Cl, 11.12. The composition of this salt is confirmed by single-crystal X-ray structure analysis.[4] The $(PF_6)^-$ salt is prepared from the chloride salt by metathesis in water using $NH_4(PF_6)$.

Properties

Crystals of $[Co(en)_2(SCH_2COO)](ClO_4)$ are dark purple, sometimes appearing black. The perchlorate salt, which crystallizes from the reaction mixture, is soluble in H_2O (1.1×10^{-2} M in 0.01 M $HClO_4$),[5] slightly soluble in dimethyl sulfoxide (DMSO), DMF, and sulfolane, and insoluble in ethanol. The chloride salt is very soluble in water and insoluble in all other solvents listed above. The $(PF_6)^-$ salt is soluble in DMSO, DMF, and tetrahydrothiophene 1,1-dioxide and insoluble in ethanol. The compound is characterized by its visible-UV spectrum,[2c] which has a peak at 518 nm ($\epsilon = 152$ M^{-1} cm^{-1}) and a characteristic sulfur-to-metal charge-transfer band at 282 nm ($\epsilon = 11,700$ M^{-1} cm^{-1}) by which the purity of the compound can be estimated.

Bis(1,2-ethanediamine)(2-mercaptoacetato-O,S)cobalt(III) salts have been used in the synthesis of a variety of complexes containing the Co$-$S bond.[2c,6]

References and Notes

1. (a) K. Hori, *Nippon Kagaku Zasshi*, **90**, 561 (1969). (b) K. Hori, *Bull. Chem. Soc. Jpn.*, **48**, 2209 (1975). (c) V. Kothari and D. H. Busch, *Inorg. Chem.*, **8**, 2276 (1969).
2. (a) L. E. Asher and E. Deutsch, *Inorg. Chem.*, **12**, 1774 (1973). (b) R. H. Lane and L. E. Bennett, *J. Am. Chem. Soc.*, **92**, 1089 (1970). (c) G. J. Kennard, Ph.D. thesis, University of Cincinnati, 1977. (d) I. Kofi Adzamli, K. Libson, J. D. Lydon, R. C. Elder, and E. Deutsch, *Inorg. Chem.*, **18**, 303 (1979). (e) M. H. Dickman, R. J. Doedens, and E. Deutsch, *Inorg. Chem.*, **19**, 945 (1980). (f) C. P. Sloan and J. H. Krueger, *Inorg. Chem.*, **14**, 1481 (1975). (g) C. A. Stein, P. E. Ellis, R. C. Elder, and E. Deutsch, *Inorg. Chem.*, **15**, 1618 (1976).
3. Concentrated perchloric acid is an oxidant, and contact with organic materials can lead to explosion or fire. A. A. Schilt, *Perchloric Acid and Perchlorates*, G. Fredrick Smith Co., Columbus, OH, 1979.

4. R. C. Elder, L. R. Florian, R. E. Lake, and A. M. Yacynych, *Inorg. Chem.*, **12**, 2690 (1973).
5. J. D. Lydon, D. L. Nosco, and E. Deutsch, unpublished observations.
6. R. C. Elder, G. J. Kennard, M. D. Payne, and E. Deutsch, *Inorg. Chem.*, **17**, 1296 (1978).
7. (a) M. Woods, J. Karbwang, J. C. Sullivan, and E. Deutsch, *Inorg. Chem.*, **15**, 1678 (1976). (b) D. L. Nosco, R. C. Elder, and E. Deutsch, *Inorg. Chem.*, **19**, 2545, 1980.
8. D. L. Nosco and E. Deutsch, *J. Am. Chem. Soc.*, **102**, 7784 (1980).
9. R. C. Elder, M. J. Heeg, and E. Deutsch, *Inorg. Chem.*, **18**, 2036 (1979).

6. TETRAKIS(BENZENETHIOLATO)METALLATE(2–) COMPLEXES, [M(SPh)$_4$]$^{2-}$, OF MANGANESE, IRON, COBALT, ZINC, AND CADMIUM AND DERIVATIVES OF THE [Fe(SPh)$_4$]$^{2-}$ COMPLEX

Submitted by D. COUCOUVANIS,* C. N. MURPHY,* E. SIMHON,* P. STREMPLE,* and M. DRAGANJAC*
Checked by ROSARIO DEL PILAR NEIRA†

The relevance of the chemistry of metal-benzenethiolate complexes derives mainly from the apparent importance of metal-cysteinyl sulfur coordination in certain metalloenzymes. The complexes presented herein belong to the class of monomeric [ML$_4$]$^{2-}$ tetrahedral complexes.

The syntheses of the metal-benzenethiolate complexes are accomplished by a ligand substitution reaction between the appropriate *O*-ethyl dithiocarbonate complexes and potassium benzenethiolate, KSPh. In the original report on the syntheses of the [M(SPh)$_4$]$^{2-}$ complexes,[1] the bis(3,4-dimercapto-cyclobutane-1,2-dione) [dithiosquaric acid] metal complexes were used in the ligand substitution reactions. The rather involved synthesis of the dithiosquarate ligand,[2] using expensive starting materials, prompted us to develop new synthetic procedures which utilize the readily available, inexpensive *O*-ethyl dithiocarbonate complexes.[3]

Among the synthetic structural analogs for the active sites in the two iron and four iron ferredoxins, Holm, Ibers, and co-workers[4] have reported on the synthesis and structural characterization of a series of complexes containing the FeS$_2$Fe and Fe$_4$S$_4$ cores, respectively.** We report new convenient procedures for the

*Department of Chemistry, the University of Iowa, Iowa City, IA 52242.
†Department of Chemistry, Case Western Reserve University, Cleveland, OH 44106.
**Editor's Note: See also synthesis 9 in this volume.

synthesis of the $[(SPh)_2FeS_2Fe(SPh)_2]^{2-}$ and $[(SPh)FeS]_4^{2-}$ members of this series which utilize the $[Fe(SPh)_4]^{2-}$ complex.

A. BIS(TETRAPHENYLPHOSPHONIUM) TETRAKIS(BENZENETHIOLATO)FERRATE(II)

$$[(C_2H_5)_4N]\,[Fe(S_2COC_2H_5)_3] + 4.5KSC_6H_5 + 2[(C_6H_5)_4P]\,Cl \longrightarrow$$

$$[(C_6H_5)_4P]_2[Fe(SC_6H_5)_4] \qquad (\text{see Ref. 5})$$

All operations are carried out under a nitrogen atmosphere using thoroughly degassed solvents. The tetraethylammonium salt of tris(O-ethyl dithiocarbonato)-ferrate(II),[3] [tris(xanthato)ferrate(II), Fe(etxant)$_3^-$], 3.8 g (6.9 mmole); potassium benzenethiolate, KSC_6H_5, 4.6 g (31 mmole), and tetraphenylphosphonium chloride, $[(C_6H_5)_4P]\,Cl$, 5.18 g (13.8 mmole) are added to 25 mL acetonitrile, CH_3CN, in a 125-mL Erlenmeyer flask. The suspension is boiled for 15 minutes and the contents of the flask are filtered while hot.

Upon cooling, the filtrate deposits large, light brown-red crystals of $[(C_6H_5)_4P]_2[Fe(SC_6H_4)_4]$. The product is washed with two 10-mL portions of absolute ethanol and subsequently with three 10-mL portions of diethyl ether. The crystals, 4.9 g, 61% yield, are dried under vacuum at room temperature (time required 45 min).

In solution, the $[Fe(SC_6H_5)_4]^{2-}$ complex anion is extremely sensitive to oxygen. Crystals of the $[(C_6H_5)_4P]^+$ salt, however, can be handled in the air for brief periods of time (ca. 1 hr). Recrystallization is not necessary, but if desired it can be carried out by dissolving the product in hot CH_3CN, filtering, and cooling to room temperature. X-Ray powder pattern: nine strongest lines at 8.25, 6.6, 4.9, 4.3, 4.15, 3.95, 3.73, 3.62, and 2.68 Å.

Anal. Calcd.: C, 73.83; H, 5.16. Found: C, 73.70; H, 5.07.

B. BIS(TETRAPHENYLPHOSPHONIUM) TETRAKIS(BENZENETHIOLATO)COBALTATE(II)

$$Co^{III}(S_2COC_2H_5)_3 + 2[(C_6H_5)_4P]\,Cl + 6KSC_6H_5 \longrightarrow [(C_6H_5)_4P]_2[Co^{II}(SC_6H_5)_4]$$

$$+ KCl + KS_2COC_2H_5 + \text{oxidized ligand by-products}$$

All operations are carried out under a nitrogen atmosphere using thoroughly degassed solvents. Tris(O-ethyl dithiocarbonato)cobalt(III)[6] (1.563 g, 3.69 mmole), 2.77 g $[(C_6H_5)_4P]\,Cl$ (7.4 mmole), and 3.29 g KSC_6H_5 (22.1 mmole) are added to 35 mL CH_3CN in a 125-mL Erlenmeyer flask. The suspension is boiled for 15 minutes, and the contents of the flask are filtered while hot. Upon cooling to room temperature, and after ca. 6 hours, the deep-green crystalline product is isolated on a sintered glass filter. Following isolation, the crystals are washed twice with 10-mL portions of absolute ethanol and twice with 10-mL portions of diethyl ether. They are then dried under vacuum. The yield of $[(C_6H_5)_4P]_2[Co^{II}(SC_6H_5)_4]$ is 1.96 g (45%). Additional product can be obtained by the addition of diethyl ether to the filtrate. Recrystallization is not necessary, but if desired it can be carried out by dissolving the product in hot CH_3CN, filtering, and cooling to ambient temperature. The X-ray powder pattern of this compound is identical to that of the analogous iron compound.

Anal. Calcd.: C, 73.64; H, 5.15. Found: C, 73.65; H, 5.19.

C. BIS(TETRAPHENYLPHOSPHONIUM) TETRAKIS(BENZENETHIOLATO)MANGANATE(II)

$$[(C_2H_5)_4N][Mn(S_2COC_2H_5)_3] + 4KSC_6H_5 + 2[(C_6H_5)_4PCl] \longrightarrow$$
$$[(C_6H_5)_4P]_2[Mn(SC_6H_5)_4] + 3KS_2COC_2H_5 + KCl + (C_2H_5)_4NCl$$

All operations are carried out under a nitrogen atmosphere using thoroughly degassed solvents. Tetraethylammonium tris(*O*-ethyl dithiocarbanato)manganate(II), 4.00 g (7 mmole), tetraphenylphosphonium chloride, 5.46 g (14 mmole), and potassium benzenethiolate, 4.93 g (33 mmole), are added to 65 mL CH_3CN in a 125-mL Erlenmeyer flask. The suspension is heated until the CH_3CN starts to boil, and refluxing is continued for 15 minutes. The hot mixture is filtered, and the dark-yellow filtrate is reduced in volume to 50 mL by evaporation of the solvent under reduced pressure. Upon standing for ca. 12 hours, the $[(C_6H_5)_4P]_2[Mn(SC_6H_5)_4]$ salt crystallizes as large yellow cubic crystals. The product is collected by filtration on a sintered-glass filter and washed twice with 10-mL portions of absolute ethanol and twice with 10-mL portions of diethyl ether. After being dried under vacuum at room temperature, 4.84 g (57% yield) of the material was obtained.

Recrystallization can be carried out as described previously for the corresponding Fe(II) and Co(II) complexes. The X-ray powder pattern of this compound is identical to that of the Co and Fe analogs.

Anal. Calcd: C, 73.89; H, 5.17. Found: C, 73.85; H, 5.20.

D. BIS(TETRAPHENYLPHOSPHONIUM) TETRAKIS(BENZENETHIOLATO)ZINCATE(II)

In a nitrogen atmosphere, 1.002 g (3.2 mmole) bis(*O*-ethyl dithiocarbonato)-

zinc(II),[7] 2.450 g (6.5 mmole) [$(C_6H_5)_4P$] Cl, and 2.10 g (14 mmole) KSC_6H_5 are added to 20 mL CH_3CN in a 125-mL Erlenmeyer flask. The suspension is brought to the boiling point of CH_3CN and boiled for 10 minutes. The mixture is then filtered hot through a medium-porosity sintered-glass filter. The bright-yellow solution is allowed to cool. Upon standing for 30 minutes, bright-yellow crystals form which are removed by filtration and washed three times with 10-mL portions of absolute ethanol and twice with 10-mL portions of diethyl ether. After drying in vacuo, the yield of [$(C_6H_5)_4P$]$_2$[Zn-$(SC_6H_5)_4$] is 2.27 g (59%). Recrystallization is not necessary, but if desired it can be carried out as described previously for the iron(II) complex. By X-ray measurements, this compound is found to be isomorphous with the other members of the series.

Anal. Calcd.: C, 73.24; H, 5.12. Found: C, 73.39; H, 5.12.

E. BIS(TETRAPHENYLPHOSPHONIUM) TETRAKIS(BENZENETHIOLATO)CADMATE(II)

This complex can be obtained from bis(O-ethyl dithiocarbonato)cadmium(II)[8] by a procedure *identical* to the one described for the [$Zn(SC_6H_5)_4$]$^{2-}$ complex, in 52% yield.

Anal. Calcd.: C, 70.45; H, 4.89. Found: C, 69.91; H, 4.82.

F. BIS(TETRAPHENYLPHOSPHONIUM) TETRAKIS(BENZENETHIOLATO) DI-μ-THIO-DIFERRATE(III)

$$2[(C_6H_5)_4P]_2[Fe(SC_6H_5)_4] + 2(C_6H_5CH_2S)_2S \longrightarrow$$
$$[(C_6H_5)_4P]_2[FeS(C_6H_5S)_2]_2 + 2(C_6H_5CH_2S)_2 +$$
$$(C_6H_5S)_2 + 2[(C_6H_5)_4P](SC_6H_5)$$

All operations are carried out under a nitrogen atmosphere using thoroughly degassed solvents. Finely ground, bis(tetraphenylphosphonium)-tetrakis (benzenethiolato)ferrate(II), [$(C_6H_5)_4P$]$_2$[Fe(SC_6H_5)_4], 2.4 g (2 mmol), is dissolved in 30 mL of warm (~50°C) N,N-dimethylformamide (DMF). To this solution, a solution of 0.57 g (2 mmol) of $(C_6H_5CH_2S)_2S$[10] in 3 mL of DMF is added with vigorous stirring. The deep violet solution thus obtained is filtered, and to the filtrate 5 mL of absolute ethanol and 30 mL of dry diethyl ether are added. The crystalline precipitate that forms, following the addition of ether, is isolated by filtering the mixture through a medium porosity sintered-glass filter and washed with two 10 mL portions of diethyl ether. After drying under vacuum, at room temperature, 1.25 g (92% yield) of pure [$(C_6H_5)_4P$]$_2$[FeS(C_6H_5S)_2]$_2$ is obtained. Time required, 25 min. The product is identical to that obtained by the synthetic procedure previously reported for this compound.[9] X-Ray powder pattern: six strongest lines at 10.0, 7.0, 5.6, 4.8, 4.45, and 3.45 Å.

Anal. Calcd.: C, 67.01; H, 4.68. Found: C, 66.48; H, 4.67.

G. BIS(TETRAPHENYLPHOSPHONIUM TETRAKIS(BENZENE-THIOLATO)TETRA-μ_3-THIO-TETRAFERRATE-(II,III)

$$2[(C_6H_5)_4P]_2[FeS(C_6H_5S)_2]_2 + Na_2S_2O_4/(18\text{-crown-6-ether}) \longrightarrow$$

$$[(C_6H_5)_4P]_2[FeS(C_6H_5S)]_4 + 2Na^+ + 2[(C_6H_5)_4P]^+ + 4C_6H_5S^-$$

All operations are carried out under a nitrogen atmosphere using thoroughly degassed solvents. A methanolic solution of $Na_2S_2O_4/(1,4,7,10,13,16,18\text{-crown-6-ether})$ is prepared as described previously.[11] A 5-mL aliquot of this solution (0.22 mmole $S_2O_4^{2-}$) is added with stirring to a solution of 0.564 g $[(C_6H_5)_4P]_2[FeS(C_6H_5S)_2]_2$ (0.44 mmole) in 30 mL DMF. The solution changes color from violet to brown-yellow and is stirred for an additional 5 minutes. Following filtration through a medium-porosity sintered-glass filter, the brown solution is diluted with 25 mL absolute ethanol, and enough diethyl ether is added until the first signs of nucleation appear on the walls of the container (200 mL diethyl ether is used). The crystalline product that forms after ca. 0.5 hour is isolated by filtration, washed with two 10-mL portions of diethyl ether, and dried under vacuum. Yield of $[(C_6H_5)_4P]_2[FeS(C_6H_5S)]_4$ is 0.30 g (95%). The crude product can be recrystallized from a DMF/ethanol/diethyl ether mixture in the same relative ratio as that outlined in the synthetic procedure above. The final yield of the recrystallized product is 73%. Time required is 1 hour. The complex is identical to an "authentic" sample obtained by the synthetic procedure reported previously.[12] X-Ray powder pattern: five strong lines at 8.0, 6.6, 6.9, 4.95, and 4.0 Å.

Anal. Calcd.: C, 58.94; H, 4.13. Found: C, 59.13; H, 4.16.

Properties

The complexes are soluble in strongly polar solvents. Single crystals of all of the $[M(SPh)_4]^{2-}$ complexes as the $[Ph_4P]^+$ salts can be obtained by the slow cooling of hot CH_3CN solutions of these salts. The structures of the isomorphous and isostructural $(Ph_4P)_2[M(SPh)_4]$ complexes (M = Mn, Fe, Co, Ni, Zn, or Cd) have been determined.[13,14] The coordination geometry of all MS_4 chromophores is distorted tetrahedral. This distortion can be described as a compression of the tetrahedron along one of the twofold axes which gives rise to an approximate

D_{2d} symmetry. The high-spin $[MS_4]^{2-}$ complexes in solution are very sensitive to oxygen and display typical electronic ligand field spectra. On the basis of the 10 Dq transitions in the near-IR spectra of these complexes, the thiophenolate ligand is placed between Cl^- and hexamethylphosphoric triamide (HMPA) in the spectrochemical series.

The $[(SPh)_2FeS_2Fe(SPh)_2]^{2-}$ and $[SPhFeS]_4{}^{2-}$ complexes originally were prepared by Holm and co-workers[9,12] who reported their syntheses and properties and established their importance as structural analogs for the two and four iron ferredoxins, respectively.

References and Notes

1. D. G. Holah and D. Coucouvanis, *J. Am. Chem. Soc.*, **97**, 6917 (1975).
2. D. Coucouvanis, D. G. Holah, and F. J. Hollander, *Inorg. Chem.*, **14**, 2657 (1975).
3. D. G. Holah and C. N. Murphy, *Can. J. Chem.*, **49**, 2726 (1971).
4. R. H. Holm and J. A. Ibers, *Iron-Sulfur Proteins*, Vol. III, W. Lovenburg (ed.), Academic Press, New York, 1977, pp. 206-281.
5. Balanced equations are difficult to write because of the unknown composition of the by-products formed in these reactions.
6. J. V. Dubsky, *J. Prakt. Chem.*, **90**, 61 (1914).
7. E. Emmet Reid, *Organic Chemistry of Bivalent Sulfur*, Vol. IV, Chemical Publishing Co., New York, 1962, Chaps. 1, 2, and 3.
8. H. M. Rietveld and E. N. Maslen, *Acta Cryst.*, **18**, 429 (1965) and references therein.
9. J. J. Mayerle, S. E. Denmark, B. V. DePamphilis, J. A. Ibers, and R. H. Holm, *J. Am. Chem. Soc.*, **97**, 1032 (1975).
10. The $(C_6H_5CH_2S)_2S$ reagent is conveniently obtained, in excellent yield, by the reaction between $C_6H_5CH_2SH$ and SCl_2 in petroleum ether in a 2:1 molar ratio.
11. T. Mincey and T. G. Traylor, *Bioinorg. Chem.*, **9**, 409 (1978).
12. B. A. Averill, T. Herskovitz, R. H. Holm, and J. A. Ibers, *J. Amer. Chem. Soc.*, **95**, 3523 (1973).
13. D. Coucouvanis, D. Swenson, N. C. Baenziger, D. G. Holah, A. Kostikas, A. Simopoulos, and V. Petrouleas, *J. Am. Chem. Soc.*, **98**, 5721 (1976).
14. D. Swenson, N. C. Baenziger, and D. Coucouvanis, *J. Am. Chem. Soc.*, **100**, 1932 (1978).

7. TRI-μ-CHLORO-CHLOROTETRAKIS(TRIPHENYLPHOS-PHINE)DIRUTHENIUM(II) COMPLEXES WITH ACETONE CARBONYL AND THIOCARBONYL LIGANDS

Submitted by P. W. ARMIT,* T. A. STEPHENSON,* and E. S. SWITKES*
Checked by ANNA TUCKA,† ROSANDA CUENCA,† JOHN F. HARROD,† and IAN S. BUTLER†

*Department of Chemistry, University of Edinburgh, Edinburgh EH9 3JJ, Scotland.
†Department of Chemistry, McGill University, Montreal, Quebec H3A 2K6, Canada.

In recent years, a number of ruthenium complexes which contain a $RuCl_3Ru$ bridging unit and a variety of terminal ligands have been synthesized. Examples include $[L_3RuCl_3RuL_3]Cl$ ($L = PR_3$, $P(OR)Ph_2$, etc.),[1] $Ru_2Cl_4(PR_3)_5$,[1] $Ru_2Cl_4(N_2)(PPh_3)_4$,[2] $Ru_2Cl_4(PF_3)(PPh_3)_4$,[3] and $[Ru_2Cl_3(CO)_2(PPh_3)_4][BPh_4]$.[4] The syntheses presented here describe the preparation of $[Ru_2Cl_4(CS)(PPh_3)_4]$-acetone and $[Ru_2Cl_4(CO)(PPh_3)_4]$-acetone (1:2), in which the key step is believed to be the intermolecular coupling of two coordinatively unsaturated (or weakly solvated) monomers.

A. TRI-μ-CHLORO-CHLORO(THIOCARBONYL)TETRAKIS(TRIPHENYL-PHOSPHINE)DIRUTHENIUM(II)-ACETONE (1:1)

$$RuCl_2(PPh_3)_3 + CS_2 \xrightarrow{\Delta} [RuCl_2(S_2CPPh_3)(PPh_3)_2]\text{-}CS_2 \; (A) +$$

$$[Ru_2Cl_4(CS)(PPh_3)_4]\text{-acetone} \; (B) + [RuCl_2(CS)(PPh_3)_2]_2$$

Procedure

Carbon disulfide (30 mL) is deoxygenated by refluxing gently in a stream of nitrogen for ca. 15 minutes. Then, to the cooled solution add 0.20 g (0.21 mmole) $RuCl_2(PPh_3)_3$ and reflux under nitrogen for ca. 10 minutes.[5] The solution is cooled in ice and the red crystalline residue A is filtered off. Yield 0.016 g (8%); mp 175-176°. This material initially contains some CS_2 of solvation [$\nu(CS_2)$ at 1515 cm^{-1}], but this can be removed by gentle suction at a water pump.

Anal. Calcd. for $C_{55}H_{45}Cl_2P_3RuS_2$‡: C, 63.8; H, 4.4% Found: C, 64.0; H, 4.2%.

Concentrate the red filtrate on a vacuum line to ca. 5-mL volume and treat the solution with an excess (30 mL) of oxygen-free light petroleum (bp 60-80°). Filter off the resulting orange-pink precipitate, wash well with light petroleum (bp 60-80°), and air dry. Triturate the product for several minutes with 10 mL of oxygen-free acetone** and filter off the crystalline red-brown complex B, wash well with acetone, and air dry. Yield 0.09 g (60%), mp 167-168°.

---

‡This compound was originally formulated as $[RuCl(\eta^2\text{-}CS_2)(PPh_3)_3]Cl$[6] but further studies on the analogous $RuCl_2(PEtPh_2)_3/CS_2$ reaction[7] (which gives much more soluble products) strongly suggests that it should be reformulated as $[RuCl_2(S_2CPPh_3)(PPh_3)_2]$ containing the $Ph_3P^+\cdot CS_2^-$ zwitterion ligand.

**It is most important to use *no more than* 10 mL of acetone in this step or else only small amounts of B are precipitated. Initially all the solid dissolves and then on *vigorous* agitation with a stirring rod, the crystalline complex B reprecipitates from solution as an acetone solvate. Small amounts of unreacted $RuCl_2(PPh_3)_3$ and the double chloride bridged complex $[\{RuCl_2CS(PPh_3)_2\}_2]$ are left in solution.

Anal. Calcd. for $C_{76}H_{66}Cl_4OP_4Ru_2S$: C, 61.0; H, 4.4; Cl, 9.5; P, 8.3; S, 2.1; M, 1494; Found: C, 60.8; H, 4.3; Cl, 9.4; P. 8.5; S, 2.3; M (in C_6H_6 by osmometry), 1423.

B. CARBONYLTRI-μ-CHLORO-CHLOROTETRAKIS(TRIPHENYLPHOS-PHINE)DIRUTHENIUM(II)-ACETONE (1:2)

$$RuCl_2(PPh_3)_3 + RuCl_2(CO)(PPh_3)_2(dmf) \xrightarrow[\text{acetone}]{\Delta} [Ru_2Cl_4(CO)(PPh_3)_4] \text{-}$$

acetone (1:2)

Procedure

Acetone (30 mL) is deoxygenated by refluxing gently in a stream of nitrogen for ca. 15 minutes. To the cooled solution, add 0.12 g (0.13 mmole) of $RuCl_2(PPh_3)_3$[5] and 0.10 g (0.13 mmole) $RuCl_2(CO)(PPh_3)_2(dmf)$[8] [dmf = N,N-dimethylformamide] and reflux under nitrogen for ca. 2.5 hours. Cool the solution in ice and filter off the deep red crystals of the product, wash well with diethyl ether and air dry. Further crystals of the complex can be obtained by evaporation of the filtrate to ca. 10 mL. Yield 0.13 g (69%), mp 170-171° (dec.).

Anal. Calcd. for $C_{79}H_{72}Cl_4O_3P_4Ru_2$: C, 61.6; H, 4.8; Cl, 9.3%; Found: C, 61.7; H, 4.7; Cl, 9.4%.

Properties

Both triple-chloro-bridged compounds are stable in air, although in solution facile oxidation occurs. They are soluble in dichloromethane, chloroform, and benzene and insoluble in acetone, diethyl ether, petroleum ether, and water. Spectral properties of the thiocarbonyl complex are: ν_{CS} 1284, ν_{CO}(acetone) 1705, ν_{RuCl} 318 (s), 308 (m), (sh), 260 (m) cm^{-1} (Nujol); ^{31}P-$\{^1H\}$ nmr spectrum (CDCl$_3$ at 298 K) 48.3 (quartet) and 36.1 (quartet) ppm, $^2J_{P1P2}$ 37.4, $^2J_{P3P4}$ 24.6, δ_{P1P2} 94.0, δ_{P3P4} 54.9 Hz.[9] Spectral properties of the carbonyl complex are: ν_{CO} 1955(s), 1939(w), ν_{CO} (acetone) 1710, ν_{RuCl} 319 (s), 250 (br) cm^{-1} (Nujol); ^{31}P-$\{^1H\}$ nmr spectrum (CDCl$_3$ at 298 K) 48.0 (quartet) and 40.3 (quartet) ppm, $^2J_{P1P2}$ 37.5, $^2J_{P3P4}$ 24.6, δ_{P1P2} 97.7, δ_{P3P4} 74.2 Hz.[4] An X-ray analysis of [Ru$_2$Cl$_4$(CS)(PPh$_3$)$_4$]-acetone confirms the structural formulation and shows Ru ... Ru 3.35 Å.[10] Reaction with other Lewis bases (L) leads to facile bridge cleavage and ligand exchange giving mixtures of RuYCl$_2$L$_3$ and RuCl$_2$L$_3$ or 4 (Y = CO, CS; L = P(OR)Ph$_2$, P(OMe)$_2$Ph).[11] Finally, electrochemical studies[12] reveal that both dimers undergo a facile, reversible, one-electron oxidation to form the 35-electron [Ru$_2$Cl$_4$Y(PPh$_3$)$_4$]$^+$ cations.

References

1. For detailed references, see P. W. Armit, A. S. F. Boyd, and T. A. Stephenson, *J. Chem. Soc., Dalton Trans.*, 1663 (1975).
2. L. W. Gosser, W. H. Knoth, and G. W. Parshall, *J. Am. Chem. Soc.*, **95**, 3436 (1973).
3. R. A. Head and J. F. Nixon, *J. Chem. Soc., Dalton Trans.*, 901 (1978).
4. P. W. Armit, W. J. Sime, and T. A. Stephenson, *J. Chem. Soc., Dalton Trans.*, 2121 (1976).
5. P. S. Hallman, T. A. Stephenson, and G. Wilkinson, *Inorg. Synth.*, **12**, 237 (1970).
6. J. D. Gilbert, M. C. Baird, and G. Wilkinson, *J. Chem. Soc., A.*, 2198 (1968).
7. P. W. Armit, W. J. Sime, T. A. Stephenson, and L. Scott, *J. Organometal. Chem.*, **161**, 391 (1978).
8. B. R. James, L. D. Markham, B. C. Hui, and G. L. Rempel, *J. Chem. Soc., Dalton Trans.*, 2247 (1973).
9. T. A. Stephenson, E. S. Switkes, and P. W. Armit, *J. Chem. Soc., Dalton Trans.*, 1134 (1974).
10. A. J. F. Fraser and R. O. Gould, *J. Chem. Soc., Dalton Trans.*, 1139 (1974).
11. W. J. Sime and T. A. Stephenson, *J. Chem. Soc., Dalton Trans.*, 1045 (1979).
12. G. A. Heath, G. Hefter, D. R. Robertson, W. J. Sime, and T. A. Stephenson, *J. Organometal. Chem.*, **152**, C1 (1978).

8. (L-CYSTEINATO)GOLD(I)

Submitted by C. FRANK SHAW III* and GERARD P. SCHMITZ*
Checked by R. C. ELDER† and CARY S. DANIEL†

$$AuBr_4^- + 3CySH \xrightarrow{HBr} CyS-SCy + Au(SCy) \downarrow + 3HBr + Br^-$$

(L-Cysteinato)gold(I) is an insoluble gold(I) thiolate which is of interest because of its possible role in the metabolism of gold(I) thiolates used in arthritis therapy.[1-4] It was first reported as a decomposition product in the reaction of chloro(triphenylphosphine)gold(I) with cysteine[5] and subsequently from the reaction of gold sodium thiomalate, $Na_{2n}[Au(SC_4H_3O_4)]_n$ [sodium [mercapto-butanedioato(3-)]aurate(I)], with cysteine[2] and by cysteine reduction of $K[AuBr_4]$ or $Na[AuCl_4]$.[3] The thiomalate reaction leads to a product which is sometimes contaminated by thiomalate.[1] The reaction with $Na[AuCl_4]$ can lead to Au(SCy) or Na(AuClSCy) depending upon the reaction conditions.[3] However, the $K[AuBr_4]$ reaction leads to nearly quantitative yields of Au(SCy) if carried

*Department of Chemistry, University of Wisconsin-Milwaukee, Milwaukee, Wisconsin, 53201.
†Department of Chemistry, University of Cincinnati, Cincinnati, OH 45221.

out in 0.1 M HBr solution.[1] The high acidity is required to prevent precipitation of the oxidation product, cystine, which has limited solubility at neutral pH.

Procedure

Potassium tetrabromoaurate(1-) (0.3005 g, 0.541mmole)[6] is placed in a 100-mL beaker containing a magnetic stirring bar. To this is added 20 mL HBr solution (0.1 M). L-Cysteine hydrochloride hydrate (available from Aldrich Chemical Co., Milwaukee, WI) (0.3120 g, 1.78mmole) is dissolved in a separate 100-mL beaker with 20 mL HBr (0.1 M). (The acid concentration is important. Use of 1 M HBr leads to a different product.) The CySH·HCl·H$_2$O solution is rapidly added to the [AuBr$_4$]$^-$ solution with constant stirring.

Decolorization of the purple [AuBr$_4$]$^-$ to yield a very pale-yellow solution occurs within 15 seconds of mixing. A white suspension is formed slowly (usually 1-5 min) and is stirred for 15 minutes. The mixture is then placed in a refrigerator overnight to allow the precipitate to coagulate. (If the precipitate is filtered at this point, the very fine particles pass through the filter and much lower yields are obtained. Alternatively, centrifugation can be substituted for filtration, in all but the final filtration from diethyl ether.) The precipitate which is formed is filtered using a sintered-glass filter. The compound is washed once with 0.1 N HBr solution, twice (25 mL) with absolute ethanol, and twice (25 mL) with diethyl ether and then dried in vacuo over P$_2$O$_5$. The yield is 90-97% (0.1682 g, 0.53mmole).

Anal. Calcd. for C$_3$H$_6$AuNO$_2$S: C, 11.36; H, 1.91; N, 4.42; Au, 62.1; Found: C, 11.32; H, 1.53; N, 4.28; Au, 64.2

Properties

(L-Cysteinato)gold(I) is a white to pale-yellow, air-stable solid that is extremely insoluble in water in the pH range 4-10: 1.1 μM at pH 7.4 and increasing slightly with increasing pH. The solubility is greater at very low (<1) or very high (>14) pH, but still not sufficient to dissolve a significant proportion of the compound. However, it can be partially redissolved in the presence of excess cysteine to form [Au(SCy)$_2$]$^-$ with a formation constant K_{eq} = [Au(SCy)$_2$]$^-$[H$^+$]/ [Au(SCy)] [CySH] of 2.1 × 10^{-3}.[1] In 5 mM cysteine at pH 7.4, 200-μM solutions of [Au(SCy)$_2$]$^-$ are obtained.[1] The complex is thermally stable and has a mp (dec.) 180-230°, but decomposes if exposed to roomlight or sunlight for long periods of time. Characteristic infrared bands, in Nujol and HCBD mulls, are observed at ν = 1680 (m), 1625 (w), 1565 (s), 1380 (s), 1350 (m,sh), 1245 (w,br), 1132 (m), 945 (w,br), 893 (w), 832 (w), 715 (vw), 655 (vw), 625 (m), and 550 (m) cm^{-1}.

Gold complexes of cysteine derivatives such as *N*-acetylcysteine, penicillamine, cysteine methyl ester, and glutathione do not appear to be as insoluble as cysteine itself.

Acknowledgment

The authors thank the checkers for several helpful suggestions that have been incorporated into the procedure.

References

1. C. F. Shaw III, G. P. Schmitz, P. Witkiewicz, and H. O. Thompson, *J. Inorg. Biochem.*, **10**, 317 (1979).
2. C. J. Danpure, *Biochem. Pharmacol.*, **25**, 2343 (1976).
3. D. H. Brown, G. C. McKinley, and W. E. Smith, *J. Chem. Soc., Dalton*, 199 (1978).
4. H. O. Thompson, J. Blaszak, C. J. Knudtson, and C. F. Shaw III, *Bioinorg. Chem.*, **9**, 375 (1978).
5. C. Kowala and J. M. Swan, *Aust. J. Chem.*, **19**, 547 (1966).
6. B. P. Block, *Inorg. Synth.*, **4**, 14 (1953).

9. TETRANUCLEAR IRON-SULFUR AND IRON-SELENIUM CLUSTERS

Submitted by GEORGE CHRISTOU,* C. DAVID GARNER,* A. BALASUBRAMANIAM,[†] BRIAN RIDGE,[†] and H. N. RYDON[†]
Checked by EDWARD I. STIEFEL[‡] and WIE-HIN PAN[‡]

$$4FeCl_3 + 14Na(Li)SR \rightarrow \frac{4}{n} [Fe(SR)_3]_n + 12Na(Li)Cl + 2Na(Li)SR$$

$$\frac{4}{n} [Fe(SR)_3]_n + 4X + 2Na(Li)SR \rightarrow Na_2(Li_2)[Fe_4X_4(SR)_4] + 5RSSR$$

$$(X = S \text{ or } Se)$$

Tetrakis(thiolato)tetra-μ_3-thio-tetrairon, $[Fe_4S_4(SR)_4]^{2-}$, complexes involve a distorted cubane-type structure with a thiolato group completing an essentially tetrahedral array of sulfur atoms about each iron. These moieties are close structural analogs of the iron-sulfur centers of oxidized four- and eight-iron

*Manchester University, Manchester M13 9PL, U.K.
[†]Exeter University, Exeter EX4 4QD, U.K.
[‡]C. F. Kettering Research Labs, Yellow Springs, OH 45387.

ferredoxin (FD_{ox}) proteins and reduced high-potential iron proteins ($HiPIP_{red}$), in which the thiolate ligation is accomplished by cysteinyl residues of the protein polypeptide.[1] The involvement of these centers as electron-transfer agents in processes so diverse and fundamental as nitrogen fixation, ATP formation, hydrogen uptake and evolution, and photosynthetic electron transport,[2] has stimulated the search for synthetic analogs in order that their properties may be better defined. Synthetic analogs of the iron-sulfur proteins have been developed successfully in a series of studies, largely by Holm et al.,[3] which have not only detailed the chemistry of these species but also stimulated and clarified the biochemical understanding of these systems.

Tetrakis(thiolato)tetra-μ_3-seleno-tetrairon, $[Fe_4Se_4(SR)_4]^{2-}$, complexes are isostructural with their sulfur counterparts and have similar chemical properties. Their involvement in biological systems is a distinct possibility but remains to be demonstrated.

$[Fe_4S_4(SR)_4]^{2-}$ complexes were first prepared by the reaction in methanol of iron(III) thiolates with a solution of sodium hydrogen sulfide and sodium methoxide (i.e., sodium sulfide was generated in situ).[4] Subsequently, a synthesis was described[5] which employed the commercially available anhydrous lithium sulfide as the sulfide source. As a further alternative, elemental sulfur, which is reduced by an excess of the thiolate anion, may be used for the in situ production of the sulfide.[6] Two syntheses of $[Fe_4Se_4(SR)_4]^{2-}$ complexes have been described. One involves the introduction of the selenium as an ethanolic solution of NaHSe, prepared by reacting equimolar amounts of elemental selenium and sodium tetrahydroborate(1-) in ethanol.[7] The other generates the selenide by reducing elemental selenium in situ with an excess of the thiolate anion.[8]

The use of the elemental chalcogenide appears to be the most convenient procedure for the synthesis of $[Fe_4X_4(SR)_4]^{2-}$ (X = S or Se) complexes in reasonable yield under ambient conditions. Preparations of representative complexes by this method are therefore described.

Procedure

The syntheses were performed on a vacuum manifold in Schlenk-type vessels under an atmosphere of nitrogen, purified by passage over a BASF R3-11 catalyst maintained at 140-150°. Solvents were Analar grade and were not distilled prior to use but were purged by four cycles of evacuation, followed by admission of nitrogen. Reagent-grade iron(III) chloride (B.D.H.), benzenethiol, 1,1-dimethylethanethiol (Koch-Light), and quaternary ammonium halides (B.D.H.) were used without further purification. Transfers of solutions and solvents were accomplished using syringes, previously flushed with nitrogen, and rubber septum caps. Mixing of reagents was assisted by the use of magnetic stirrers.

There are no particularly hazardous steps in the procedures described below, although care must be exercised when handling metallic sodium or lithium.

■ **Caution.** *It is recommended that as many of the manipulations as possible be carried out in a hood with a good exhaust. This is particularly important with respect to the volatile mercaptans which can induce allergic reactions in some people; any excess mercaptan may be destroyed by the addition of an oxidizing agent such as Clorox. Selenium-containing materials have been noted to have potentially adverse health effects.*[9]

A. $(Bu_4N)_2[Fe_4S_4(SPh)_4]$

Sodium (0.92 g, 40 mmole) or lithium (0.28 g, 40 mmole) is carefully dissolved in methanol (40 mL) while cooling the flask to $\leqslant 5°$; and, after the flask and its contents have attained room temperature, benzenethiol (4.1 mL, 40 mmole) is added. Anhydrous iron(III) chloride (1.62 g, 10 mmole) dissolved in methanol (25 mL) is added to this solution and a deep yellow-brown coloration is obtained after a few moments. Elemental sulfur (0.32 g, 10 mmole) is now added. The elemental chalcogen is conveniently added by weighing it into a small glass vial and adding this to the reaction mixture by momentarily removing the septum cap. After replacement of the septum cap, the flask should be quickly evacuated and then flushed with nitrogen to ensure oxygen-free conditions. After ca. 5 minutes the solution begins to acquire the deep red-brown color characteristic of $[Fe_4S_4(SPh)_4]^{2-}$. The solution is stirred for ca. 12 hours at room temperature and then filtered into a solution of tetrabutylammonium iodide (2.77 g, 7.5 mmole) in methanol (20 mL), whereupon a fine, black powder is precipitated. This material is collected by filtration, washed copiously with methanol, and dried in vacuo. Purification is achieved by dissolving the powder in the minimum volume (ca. 35 mL) of warm (45-50°) acetonitrile, filtering, and then adding methanol (ca. 35 mL) to incipient crystallization, while maintaining the temperature at 45-50°. Slow cooling to room temperature and then standing at $-5°$ overnight produces the compound as large, black needle-shaped crystals. These are collected by filtration, washed with methanol, and dried in vacuo. Yield* 2.2-2.7 g (70-85%).

Anal. Calcd. for $C_{56}H_{92}Fe_4N_2S_8$: C, 52.8; H, 7.3; Fe, 17.5; N, 2.2; S, 20.1. Found: C, 52.5; H, 7.2; Fe, 17.6; N, 2.1; S, 20.2; mp 190-191° (sealed tube under nitrogen).

Essentially the same procedure may be used for the synthesis and isolation of other tetralkylammonium salts of $[Fe_4X_4(SR)_4]^{2-}$ (X = S or Se; R = alkyl or aryl) derivatives. Details for the preparation of three such compounds are summarised below.

B. (Me₄N)₂[Fe₄S₄(S-*t*-Bu)₄]

This compound is obtained by the procedure for Section 9-A, using sodium (1.38 g, 60 mmole), 1,1-dimethylethanethiol (6.68 mL, 60 mmole), iron(III) chloride (2.43 g, 15 mmole), elemental sulfur (0.48 g, 15 mmole), and tetramethylammonium bromide (1.54 g, 10 mmole); 2.2-2.6 g (70-80%) of the crude material is obtained if, after filtration into tetramethylammonium bromide, the reaction solution is allowed to stand at −5° overnight. Black needle-like crystals of the monoacetonitrile adduct can be obtained by adding ethyl acetate to a filtered solution of the compound in warm (45-50°) acetonitrile to incipient crystallization, followed by slow cooling and standing overnight at ca. −5°. Yield 40-50%.

Anal. Calcd. for $C_{24}H_{60}Fe_4N_2S_8 \cdot CH_3CN$: C, 34.8; H, 7.1; Fe, 24.9, N, 4.7; S, 28.6. Found: C, 34.8; H, 7.0; Fe, 25.0; N, 4.6; S, 28.1; mp 152-154° (dec.) (sealed tube under nitrogen).

C. (Bu₄N)₂[Fe₄Se₄(SPh)₄]

■ **Caution.** *Selenium compounds should be handled with extreme caution due to their potentially adverse biological effects.*[9]

A procedure similar to that described for Section 9-A may be used to prepare this compound from lithium (0.28 g, 40 mmole), benzenethiol (4.1 mL, 40 mmole), iron(III) chloride (1.62 g, 10 mmole), selenium powder (0.79 g, 10 mmole), and tetrabutylammonium iodide (2.77 g, 7.5 mmole). The black precipitate should be collected by filtration, washed copiously with methanol, and dried in vacuo. This solid can be recrystallized by dissolution in the minimum volume of acetonitrile at 45°, adding methanol to incipient crystallization at this temperature and allowing the solution to cool slowly to room temperature before leaving it overnight at ca. −5°. The compound (2.9 g, 80%)* crystallizes as large, black prisms.

Anal. Calcd. for $C_{56}H_{92}Fe_4N_2S_4Se_4$: C, 46.0; H, 6.4; Fe, 15.3; N, 1.9; S, 8.8; Se, 21.6. Found: C, 46.2; H, 6.2; Fe, 15.5; N, 2.1; S, 8.8; Se, 21.5; mp 220-221° (dec.) (sealed tube under nitrogen).

---

*The yields reported are obtained only with the minimum volume of MeCN. Consequently, more dilute acetonitrile filtrates should be concentrated in vacuo prior to MeOH addition.

D. $(Bu_4N)_2[Fe_4Se_4(S\text{-}t\text{-}Bu)_4]$

■ **Caution.** *Selenium compounds should be handled with extreme caution due to their potentially adverse biological effects.*[9]

A procedure similar to that described for Section 9-A may be used to prepare this compound from lithium (0.56 g, 80 mmole), 1,1-dimethylethanethiol (8.90 mL, 80 mmole), iron(III) chloride (3.24 g, 20 mmole), selenium powder (1.58 g, 20 mmole), and tetrabutylammonium iodide (5.54 g, 15 mmole). The fine, black precipitate should be collected by filtration, washed copiously with methanol, and dried in vacuo. Recrystallization is achieved by dissolving the compound in the minimum of warm acetonitrile ($\leqslant500$ mL), filtering, and then adding ethyl acetate (ca. 170 mL) to incipient crystallization, followed by slow cooling to room temperature before leaving at ca. $-10°$ overnight. The compound (3.60 g, 52%) crystallizes as small, black plates.

Anal. Calcd. for $C_{48}H_{108}Fe_4N_2S_4Se_4$: C, 41.7; H, 7.9; Fe, 16.2; N, 2.0; S, 9.3; Se, 22.9. Found: C, 41.9; H, 7.9; Fe, 16.7; N, 1.9; S, 9.3; Se, 22.8; mp 198-201° (dec.) (sealed tube under nitrogen).

References

1. R. H. Holm and J. A. Ibers, *Iron-Sulfur Proteins*, Vol. III, W. Lovenberg (ed.), Academic Press, New York, 1977.
2. D. O. Hall, K. K. Rao, and R. Cammack, *Sci. Prog. Oxf.*, **62**, 285 (1975).
3. R. H. Holm, *Acc. Chem. Res.*, **10**, 427 (1977), and references therein.
4. B. A. Averill, T. Herskovitz, R. H. Holm, and J. A. Ibers, *J. Am. Chem. Soc.*, **95**, 3523 (1973).
5. G. N. Schrauzer, G. W. Kiefer, K. Tano, and P. A. Doemeny, *J. Am. Chem. Soc.*, **96**, 641 (1974).
6. G. Christou and C. D. Garner, *J. Chem. Soc., Dalton Trans.*, 1093 (1979).
7. M. A. Bobrik, E. J. Laskowski, R. W. Johnson, W. O. Gillum, J. M. Berg, K. O. Hodgson, and R. H. Holm, *Inorg. Chem.*, **17**, 1402 (1978).
8. G. Christou, B. Ridge, and H. N. Rydon, *J. Chem. Soc., Dalton Trans.*, 1423 (1978).
9. D. L. Klayman and W. H. H. Gunther (eds.), *Organic Selenium Compounds: Their Chemistry and Biology*, Wiley-Interscience, New York, 1975.

10. POLYNUCLEAR CYCLOPENTADIENYLIRON COMPLEXES CONTAINING THE S_2 LIGAND

Submitted by G. J. KUBAS* and P. J. VERGAMINI*
Checked by M. RAKOWSKI DUBOIS† and DOUGLAS J. MILLER†

*Los Alamos National Laboratory, University of California, Los Alamos, New Mexico 87545.
†Department of Chemistry, University of Colorado, Boulder, CO 80309.

In recent years, sulfur, both in its atomic and molecular (e.g., S_2) forms, has been noted to be a remarkably versatile ligand in transition metal complexes.[1,2] It has been found to bridge two, three, or even four metal atoms and thus readily lends itself to the study of polynuclear complexes. One of the classic examples is $[CpFeS]_4$ ($Cp = \eta^5\text{-}C_5H_5$), a cubane cluster containing triply bridging sulfur atoms.[3] An important property of these cluster compounds is their general tendency to undergo reversible redox reactions via modification of metal-metal interactions. For example, $[CpFeS]_4$ can exist in several different molecular oxidation states.[4]

The compounds whose syntheses are described below all contain CpFe moieties bridged by sulfur ligands, the most interesting of which is S_2. Each of the compounds is ultimately derived from commercially available* $[CpFe(CO)_2]_2$. In fact, all of the neutral S_2-containing clusters reported here were originally found to be among the products of the reaction of $[CpFe(CO)_2]_2$ with an ethyl polysulfide mixture of nominal stoichiometry $Et_2S_{3.3}$.[5] However, the yields were quite low, and more rational, higher-yield preparations were eventually developed[5]. $Cp_2Fe_2(S_2)(SEt)_2$, which contains a coplanar Fe−S−S−Fe unit,[6] can be prepared by two different methods. The reaction of S_8 with $[CpFe(CO)(SEt)]_2$ is relatively convenient, but much better yields are obtained by reacting ammonium polysulfide with $[CpFe(NCCH_3)(SEt)]_2[PF_6]_2$.[5] The latter, whose synthesis is included below, is a useful starting material for the preparation of other dinuclear mercaptide-bridged iron clusters as well. The synthesis of the cubane-like cluster $Cp_4Fe_4S_6$ is similar to that reported for $[CpFeS]_4$,[3a] in which $Cp_4Fe_4S_6$ is undoubtedly a minor reaction product also. (In the preparation of $[CpFeS]_4$ given in Reference 3a, the crude reaction product was extracted with hot bromobenzene. It was found that both $Cp_4Fe_4S_6$ and $Cp_4Fe_4S_5$ are rapidly converted to $[CpFeS]_4$ in refluxing bromobenzene.)

■ **Caution.** *The syntheses described below should be carried out in a well-ventilated hood because of the release of CO and the strong stench of mercaptides. Schlenk-type glassware should be used for most procedures.*

---

*$[CpFe(CO)_2]_2$ was purchased from Strem Chemicals, Inc., Danvers, MA 01923, or Pressure Chemical Co., Pittsburgh, PA 15201. The purity was found to be somewhat variable and, in most cases, several percent insoluble material was present. The latter can be readily removed by recrystallization of the $[CpFe(CO)_2]_2$ from CH_2Cl_2-hexane under nitrogen. However, in all cases the syntheses can be successfully carried out using the unpurified reagent.

A. BIS(ACETONITRILE)BIS(η^5-CYCLOPENTADIENYL)BIS-μ-(ETHANETHIOLATO)DIIRON(2+) BIS(HEXAFLUOROPHOSPHATE)

$$[CpFe(CO)_2]_2 + Et_2S_2 \xrightarrow[\text{reflux}]{\text{methylcyclohexane}} [CpFe(CO)(SEt)]_2 + 2CO$$

$$[CpFe(CO)(SEt)]_2 + 2[NH_4][PF_6] + \frac{1}{2}O_2 \xrightarrow{CH_3CN}$$

$$[CpFe(NCCH_3)(SEt)]_2[PF_6]_2 + 2NH_3 + H_2O + 2CO$$

Procedure

Before proceeding, it should be noted that the synthesis detailed below will require at least one week. A less time-consuming, though more complex, scheme is described in a footnote.

A mixture of $[CpFe(CO)_2]_2$ (24.7 g, 70 mmole), Et_2S_2 (Eastman Organic Chemicals, Rochester, NY 14650) (42 ml, 344 mmole), and methylcyclohexane (300 ml) is refluxed under nitrogen in a 500-mL, round-bottomed flask for 16 hours. The reaction mixture is cooled and quickly filtered through a large medium-porosity frit into a 500-mL filter flask (■ *Caution. The precipitate collected on the frit is pyrophoric. The filtration can be done in air, but the precipitate must not be allowed to become dry in the air stream or spontaneous combustion may result. The frit should be filled with water as soon as the filtration is finished.*) The filtrate is then transferred to a 500-mL, round-bottomed flask (exclusion of air is unnecessary), and all volatiles, including unreacted Et_2S_2, are removed by rotoevaporation at $\leqslant 0.1$ torr. The residue of crude $[CpFe(CO)(SEt)]_2$ is placed into a large polyethylene beaker or other nonglass container (fluoride attack on glass will lead to impure product) along with $(NH_4)(PF_6)$ (99.5% purity; purchased from Ozark-Mahoning Co., Tulsa, OK 74119) (50 g, 307 mmole) and 750 mL CH_3CN. The resulting solution is then stirred magnetically with good exposure to atmospheric oxygen (e.g., bubbling air through the solution) until the color of the solution becomes deep blood-red by transmitted light. At ambient temperature, a week or more may be necessary for this to occur, although smaller-scale preparations may take less time.* The

*During this time, $[CpFe(CO)(SEt)]_2$ is oxidized stepwise to its dication, which loses CO to give $[CpFe(NCCH_3)(SEt)]_2{}^{2+}$. Ammonia is evolved as a result of proton removal from $NH_4{}^+$. The reaction is complex in that pure samples of $[CpFe(CO)(SEt)]_2[PF_6]_2$ prepared by other methods of oxidation[9] lose CO extremely slowly in CH_3CN and in a stepwise manner (i.e., $[CpFe(CO)(NCCH_3)(SEt)]_2[PF_6]_2$ can be isolated as an intermediate). However, it has been found that $[CpFe(CO)(SEt)]_2[PF_6]_2$ (0.71 g) can be converted to $[CpFe(NCCH_3)(SEt)]_2[PF_6]_2$ in 70% yield by a 4-hour reflux in CH_3CN (25 mL) in the presence of excess $[NH_4][PF_6]$ (0.5 g) and atmospheric oxygen. Thus, if time is a crucial factor, it is recommended that the crude $[CpFe(CO)(SEt)]_2$ be oxidized to the PF_6 salt of its dication with bromine,[9] followed by conversion to the acetonitrile complex by the reflux procedure.

reaction rate can be increased somewhat by warming the solution to 34-40° (excessive decomposition results at higher temperatures). In any event, the solution volume should be maintained at about 500 mL or greater by replenishment with CH_3CN.

The color of the reaction mixture will initially change to deep blue-green, then to deep green, and finally to red. The volume of the reaction mixture is then reduced to 100 mL by rotoevaporation, followed by slow addition of 650 mL H_2O with stirring. The resulting brown precipitate is collected on a medium-porosity frit, washed thoroughly with water, and dried. Extraction of the crude product through the frit with CH_3CN (650-750 mL) followed by reduction of filtate volume to 100 mL and addition of CH_2Cl_2 (400 mL) gave 25.7 g (50% yield based on $[CpFe(CO)_2]_2$) of brown-black, crystalline $[CpFe(NCCH_3)(SEt)]_2[PF_6]_2$, which was washed with CH_2Cl_2 and air dried.

Anal. Calcd. for $C_{18}H_{26}N_2S_2Fe_2P_2F_{12}$: C, 29.4; H, 3.6; N, 3.8; S, 8.7; Fe, 15.2; P, 8.4; F, 31.0. Found: C, 29.3; H, 3.6; N, 3.8; S, 8.7; Fe, 15.0; P, 8.5; F, 30.8.

An SMe analog (see Ref. 5; An SPh analogue could not be prepared) can be prepared in a similar fashion. The time required for the oxidation step is significantly less than that for the SEt complex, and care must be taken that the product is isolated as soon as possible since $[CpFe(NCCH_3)(SMe)]_2[PF_6]_2$ is somewhat unstable in the reaction mixtures, slowly decomposing to a brown insoluble solid.

Properties

The compound $[CpFe(NCCH_3)(SEt)]_2[PF_6]_2$ is a nearly black, crystalline solid, moderately soluble in CH_3CN giving air-stable, deep-red solutions. It is insoluble in water, CH_2Cl_2, and nonpolar solvents and should be recrystallized from and characterized in CH_3CN since the CH_3CN ligands are labile and readily replaceable by other donor solvents as well as by charged ligands such as sulfide, cyanide, and thiocyanate.[2] The complex is diamagnetic and gives sharp proton nmr resonances in CD_3CN at τ = 4.62 s[Cp], 7.51 q[CH_2], 8.25 t[CH_3], and 7.90 s[CH_3CN]. The coordinated CH_3CN gives an infrared absorption at 2299 cm^{-1}. An X-ray structural determination confirmed the dimeric nature of the complex and the presence of terminally bound CH_3CN and bridging mercaptides.[2]

B. BIS(η^5-CYCLOPENTADIENYL)-μ-(DISULFUR) BIS-μ-(ETHANE-THIOLATO)-DIIRON (METHOD I)

$$[CpFe(CO)_2]_2 + Et_2S_2 \xrightarrow[\text{reflux}]{\text{methylcyclohexane}} [CpFe(CO)(SEt)]_2 + 2CO$$

$$[\text{CpFe(CO)(SEt)}]_2 + \frac{1}{4}\,S_8 \xrightarrow[\text{reflux}]{\text{methylcyclohexane}} Cp_2Fe_2(S_2)(SEt)_2 + 2CO$$

Procedure

A solution of $[\text{CpFe(CO)(SEt)}]_2$ is prepared in exactly the same manner as described in Section 10-A except that it is filtered under nitrogen into a 500-mL, round-bottomed flask. Powdered sulfur (13.2 g, 51 mmole) is then added and the magnetically stirred mixture is refluxed for 3.5 hours. After being cooled to about 40°, the reaction mixture is filtered through a medium-porosity frit and poured onto a hexane-washed alumina column (4 × 45 cm). Elution with nitrogen-saturated benzene gives an initial yellow band of ferrocene, which is discarded, closely followed by a deep-blue band of $Cp_2Fe_2(S_2)(SEt)_2$. The benzene eluate containing the latter is collected in a 500-mL, round-bottomed flask with minimal atmospheric exposure, and solvent is removed by rotoevaporation. The resulting solid residue is dissolved in 25 mL deoxygenated CS_2 and filtered through a medium-porosity frit into a 100-mL flask. After partial solvent removal in vacuo, 35 mL deoxygenated ethanol is added, and the solution volume is further reduced to 25-30 mL. The resulting black, crystalline precipitate of $Cp_2Fe_2(S_2)(SEt)_2$ is collected on a coarse frit, washed with a small amount of hexane, and air dried. The product is sufficiently pure for most purposes and weighs 2.3 g (8% yield based on $[\text{CpFe(CO)}_2]_2$).

Anal. Calcd. for $C_{14}H_{20}S_4Fe_2$: C, 39.3; H, 4.7; S, 29.9; Fe, 26.1. Found: C, 39.4; H, 4.8; S, 29.9; Fe, 26.0.

An SMe, but not an SPh, analog can be prepared by a procedure similar to the above.

C. BIS(η^5-CYCLOPENTADIENYL)-μ-(DISULFUR) BIS-μ-(ETHANE-THIOLATO)-DIIRON (METHOD II)

$$[\text{CpFe(NCCH}_3)\text{(SEt)}]_2[\text{PF}_6]_2 + S_2^{2-}(\text{aq}) \xrightarrow{\text{CH}_3\text{CN}} Cp_2Fe_2(S_2)(SEt)_2 +$$
$$2[\text{PF}_6]^-$$

Procedure

Under inert-atmosphere conditions, a mixture of 20 mL ammonium polysulfide solution ("20% $(NH_4)_2S$" purchased from J. T. Baker Chemical Co., Phillipsburg, NJ 08865)* and 20 mL methanol is added dropwise from a separatory

---

*Aqueous sodium sulfide containing dissolved sulfur ($\frac{1}{8}$ mole S_8 per mole Na_2S) can also be used. The yield of product was found to be 61% in this case.

funnel over a period of 10-15 minutes to a 500-mL, round-bottomed flask containing a vigorously stirred solution of $[CpFe(NCCH_3)(SEt)]_2[PF_6]_2$ (7.36 g, 10 mmole) in CH_3CN (300 mL). The reaction mixture is stirred an additional 30 minutes, during which time it becomes grey-green in color. (A brown precipitate, presumably a decomposition product, also forms.) After all solvent is removed by rotoevaporation, deoxygenated CS_2 (150 mL) is added to the residue, and the mixture is stirred thoroughly for several minutes. The extract is then filtered through a medium-porosity frit into a 250-mL flask and rotoevaporated to small volume (40-50 mL). The concentrated solution is chromatographed on a 4 × 20 cm alumina column with minimal atmospheric exposure. The $Cp_2Fe_2(S_2)(SEt)_2$ is eluted with benzene and isolated as described in Method I. The yield is 2.7 g (63%).

Properties

The compound $Cp_2Fe_2(S_2)(SEt)_2$ is a shiny black, crystalline solid readily soluble in CS_2, CH_2Cl_2, or benzene to give deep grey-green solutions which slowly decompose in air. It is slightly soluble in nonpolar solvents such as hexane and gives a sharp singlet pmr resonance due to Cp at τ 5.27 in CS_2. The ethyl resonance occurs as an unresolved multiplet at τ 9.87. The infrared spectrum shows a band due to $\nu(S-S)$ at 507 cm^{-1}. The complex can be reversibly oxidized electrochemically to a paramagnetic monocation, but further oxidation results in loss of the S_2 ligand and formation of $[CpFe(L)(SEt)]_2^{2+}$ (L = solvent).[7,8] Thus, $Cp_2Fe_2(S_2)(SEt)_2$ and $[CpFe(NCCH_3)(SEt)]_2[PF_6]_2$ are readily interconvertible.

D. TETRAKIS(η^5-CYCLOPENTADIENYL)BIS-μ_3-(DISULFUR) DI-μ_3-THIO-TETRAIRON

$$2[CpFe(CO)_2]_2 + \frac{3}{4}\,S_8 \xrightarrow[\text{reflux}]{\text{toluene}} Cp_4Fe_4S_6 + 8CO$$

Procedure

A mixture of $[CpFe(CO)_2]_2$ (10 g, 28 mmole), S_8 (5 g, 19 mmole), and toluene (175 mL) is heated at reflux under nitrogen in a 500-mL flask for 3 hours with vigorous magnetic stirring. (■ **Caution:** *Foaming due to rapid CO loss occurs during initial reflux.*) The reaction mixture is cooled and filtered through a medium-porosity frit (inert atmosphere unnecessary). The black precipitate collected on the frit is washed with toluene (50 mL), dried in air, and placed in a 250-mL flask. Dichloromethane (100 mL) is added, and the mixture is stirred for 15 minutes under nitrogen. The extract is filtered through a medium-porosity frit into a 250-mL flask, then saturated with SO_2 gas, and allowed to stand for 1 hour under an SO_2 atmosphere. A crystalline, black precipitate of $Cp_4Fe_4S_6 \cdot 2SO_2 \cdot \frac{1}{4}CH_2Cl_2$ (3.39 g, 29%) deposits, allowing separation of $Cp_4Fe_4S_6$ from $Cp_4Fe_4S_4$, a coproduct of the reaction. (If SO_2 is not available, most of the less soluble $Cp_4Fe_4S_4$ impurity can be removed by fractional crystallization. However, the $Cp_4Fe_4S_6$ purified in this manner will be somewhat less pure than that isolated via the SO_2 adduct.) The SO_2 adduct is washed with diethyl ether, air dried, and redissolved under nitrogen in 200 mL $CHCl_3$ at $\sim40°$ in a 500-mL flask. The SO_2, which is weakly coordinated to the basic sulfur atoms, is driven off by alternately briefly removing volatile substances in vacuo and reheating the solution to $\sim40°$ for several minutes. After several such cycles, the solution is filtered through a medium frit, and the volume of the filtrate is reduced to 50-75 mL in vacuo. Heptane (120 mL) is added, the solution volume is reduced to 150 mL, and the resultant precipitate is filtered off, washed with diethyl ether, and air dried. The yield of $Cp_4Fe_4S_6 \cdot xCHCl_3$ ($x < 1$) is 2.66 g (ca. 25%). This product contains a small percentage of insoluble material which can be removed by recrystallization from CH_2Cl_2 (100 mL). Addition of heptane (100 mL) to the filtered solution (N_2 atmosphere) followed by reduction of solution volume to ~125 mL, yields 2.01 g of microcrystalline $Cp_4Fe_4S_6 \cdot \frac{1}{4}CH_2Cl_2$.

Anal. Calcd. for $C_{20.25}H_{20.5}Fe_4S_6Cl_{0.5}$: C, 34.9; H, 3.0; Fe, 32.0; S, 27.6; Cl, 2.5. Found: C, 35.3; H, 3.2; Fe, 31.8; S, 26.8; Cl, 2.6.

Properties

The cluster $Cp_4Fe_4S_6$ is air stable in the solid state, but solutions slowly decompose in air. It is moderately soluble in halogenated solvents and slightly soluble in benzene, toluene, CS_2, acetone, and CH_3CN. The complex is somewhat thermally unstable and should be stored in a freezer compartment of a refrigerator. (It has been found that $Cp_4Fe_4S_6$ is converted to $[CpFeS]_4$ in less than 15 minutes in refluxing bromobenzene[5].) The proton nmr of $CDCl_3$ solutions of the complex shows sharp singlet Cp resonances of equal intensity at τ 5.18 and 5.67. Infrared peaks due to $\nu(S-S)$ occur at 497 and 509 cm^{-1}.

The structure[2,5] of $Cp_4Fe_4S_6$ has been determined and reveals a cubane-type arrangement of CpFe moieties, two triply bridging sulfur atoms, and two triply bridging S_2 groups. The latter are arranged in such a fashion that one sulfur atom from each S_2 is capable of coordinating to other metals, thus enabling the cluster to serve as a bidentate ligand. Examples of such behavior include $[(Cp_4Fe_4S_6)_2Ag][SbF_6]_3$[2,5] and $[Mo(CO)_4(Cp_4Fe_4S_6)]$.[9] $Cp_4Fe_4S_6$ is air oxidized in CH_3CN containing $[NH_4][PF_6]$, with concomitant loss of a sulfur atom to form $[Cp_4Fe_4S_5][PF_6]_2$,[5] as will be shown below.

E. TETRAKIS(η^5-CYCLOPENTADIENYL)-μ_3-(DISULFUR)TRI-μ_3-THIO-TETRAIRON(2+) HEXAFLUOROPHOSPHATE

$$Cp_4Fe_4S_6 + 1/2\ O_2 + 2[NH_4][PF_6] \xrightarrow{CH_3CN} [Cp_4Fe_4S_5][PF_6]_2 \cdot CH_3CN +$$

$$\frac{1}{8}S_8 + 2NH_3 + H_2O$$

Procedure

A mixture of $Cp_4Fe_4S_6$ (1.0 g, 1.48 mmole), $[NH_4][PF_6]$ (2.0 g, 12.3 mmole), and CH_3CN (150 mL) is stirred in a polyethylene beaker exposed to the atmosphere for 3 days. The resulting deep-brown solution is filtered and rotoevaporated in a 250-mL flask to a volume of 25 mL. Dichloromethane (100 mL) is added, and the resulting brown precipitate is collected on a medium-porosity frit, washed with CH_2Cl_2, air dried, and then washed thoroughly with water, ethanol, and hexane. The complex is extracted through the frit with several portions of CH_3CN (total volume = 60 mL) into a filter flask, and the filtrate is reduced in volume to 25 mL and treated with 100 mL CH_2Cl_2. The microcrystalline precipitate of $[Cp_4Fe_4S_5][PF_6]_2 \cdot CH_3CN$ is collected on a coarse frit, washed with CH_2Cl_2, and dried in vacuo. The yield is 0.87 g (63%).

Anal. Calcd. for $C_{22}H_{23}NS_5Fe_4P_2F_{12}$: C, 27.1; H, 2.4; Fe, 22.9; S, 16.4; P, 6.4. Found: C, 26.6; H, 2.4; Fe, 23.4; S, 16.4; P, 6.4.

Properties

The dicationic salt is an air-stable, brown solid soluble in polar solvents and insoluble in CH_2Cl_2 and nonpolar solvents. Its infrared spectrum indicated the presence of CH_3CN^* in the crystal lattice ($\nu_{CH_3CN} = 2255$ cm^{-1}) and the S_2 ligand ($\nu_{S-S} = 466$ cm^{-1}). The dication is diamagnetic and in CD_3CN gives sharp singlet pmr resonances in a 2:1:1 ratio due to Cp at τ 4.15, 4.20, and 4.83.[†] It undoubtedly possesses a cubane-like structure and can be reversibly reduced or oxidized to $[Cp_4Fe_4S_5]^n$ ($n = 0, +1,$ or $+3$).[5]

---

*The checkers did not observe an infrared peak due to CH_3CN, but we have confirmed the presence of CH_3CN by pmr spectra of the complex in DMSO-d_6.

[†]In some cases, the two low-field resonances cannot be resolved. This is most probably due to the presence of paramagnetic impurities. Also, above 70° these signals coalesce to a sharp singlet due to a fluxional process.[5]

F. TETRAKIS(η^5-CYCLOPENTADIENYL)-μ_3-(DISULFUR)TRI-μ_3-THIO-TETRAIRON

$$[Cp_4Fe_4S_5][PF_6]_2 \cdot CH_3CN \xrightarrow[CH_3CN-H_2O]{Na[BH_4]} Cp_4Fe_4S_5$$

Procedure

The complex prepared in Section 10-E (2.5 g, 2.56 mmole) is dissolved in 150 mL CH_3CN in a 250-mL flask, and H_2O (50 mL) is added. The solution is deoxygenated, cooled to $0°$, and stirred while a deoxygenated solution of Na[BH$_4$] (0.57 g, 15 mmole) in H_2O (25 mL) is added dropwise over a 15-minute period. The reaction mixture is then stirred for 1 hour under nitrogen, followed by collection of the black precipitate on a medium-porosity frit (inert atmosphere unnecessary). The crude product (1.61 g) is washed with water, ethanol, and acetone and then is recrystallized under nitrogen from CH_2Cl_2 (120 mL). The CH_2Cl_2 solution is filtered, reduced in volume to 20 mL, and treated with hexane (40 mL). The platelike crystalline aggregates of $Cp_4Fe_4S_5 \cdot 0.5CH_2Cl_2$ are collected on a coarse frit, washed with hexane, and dried in vacuo. Yield 1.454 g, 83%.

Anal. Calcd. for $C_{20.5}H_{21}S_5Fe_4Cl$: C, 35.9; H, 3.1; Fe, 32.5; S, 23.3; Cl, 5.2. Found: C, 35.9, H, 3.0; Fe, 31.7; S, 22.6; Cl, 4.9.

Properties

The cluster is a black, crystalline solid with properties similar to those of $Cp_4Fe_4S_6$. Chloroform-d_3 solutions give a temperature-dependent pmr signal (broadened singlet Cp resonances in a 3:1 ratio at τ 5.57 and 5.12 at $35°$ developing into a 1:2:1 pattern at $-40°$).[5] An infrared peak assignable to ν(S—S) occurs at 525 cm^{-1}. The structure of $Cp_4Fe_4S_5$ is expected to be similar to that of $Cp_4Fe_4S_6$ except that only one μ_3-S_2 ligand is present. Like $Cp_4Fe_4S_6$, $Cp_4Fe_4S_5$ rapidly loses sulfur in refluxing bromobenzene to give [CpFeS]$_4$ and also forms an SO_2 adduct.

Acknowledgement

This work performed under the auspices of the U.S. Department of Energy, Office of Energy Research.

References

1. (a) H. Vahrenkamp, *Angew. Chem. Int. Ed.*, **14**, 322 (1975). (b) A. Müller, *Inorg. Chem.*, **18**, 2631 (1979).
2. P. J. Vergamini and G. J. Kubas, *Prog. Inorg. Chem.*, **21**, 261 (1976).
3. (a) R. A. Schunn, C. J. Fritchie, Jr., and C. T. Prewitt, *Inorg. Chem.*, **5**, 892 (1966). (b) C. H. Wei, G. R. Wilkies, P. M. Treichel, and L. F. Dahl, *Ibid.*, **5**, 900 (1966).
4. Trinh-Toan, B. K. Teo, J. A. Ferguson, T. J. Meyer, and L. F. Dahl, *J. Am. Chem. Soc.*, **99**, 408 (1977), and references therein.
5. G. J. Kubas and P. J. Vergamini, *Inorg. Chem.* **20**, 2667 (1981).
6. (a) G. J. Kubas, T. G. Spiro, and A. Terzis, *J. Am. Chem. Soc.*, **95**, 273 (1973). (b) A. Terzis and R. Rivest, *Inorg. Chem.*, **12**, 2132 (1973).
7. G. J. Kubas, P. J. Vergamini, M. P. Eastman, and K. B. Prater, *J. Organometal. Chem.*, **117**, 71 (1976).
8. P. J. Vergamini, R. R. Ryan, and G. J. Kubas, *J. Am. Chem. Soc.*, **98**, 1980 (1976).
9. J. A. deBeer, R. J. Haines, R. Greatrex, and J. A. van Wyk, *J. Chem. Soc., Dalton*, 2341 (1973).

Chapter Two

DINUCLEAR AND POLYNUCLEAR COMPOUNDS

11. BINUCLEAR TRANSITION METAL COMPLEXES BRIDGED BY METHYLENEBIS(DIPHENYLPHOSPHINE)

Submitted by ALAN L. BALCH* and LINDA S. BENNER*
Checked by JOHN D. BASIL[†]

■ **Caution.** *Phosphines such as those used here should be handled with extreme care. Use only in a well ventilated hood!*

Methylene bis(diphenylphosphine) [bis(diphenylphosphino)methane] is a useful ligand for constructing binuclear metal complexes with novel chemical properties. The flexibility of this ligand allows the metal-metal separations in trans-bridged, binuclear species to vary from 2.1 to 3.5 Å.[1] Mononuclear complexes in which bis(diphenylphosphino)methane acts as a chelating ligand are also known but are less common. The synthesis of $Pd_2(Ph_2PCH_2PPh_2)_2Cl_2$ reported here is more reliable than the original synthesis[2] which starts with $[Pd(CO)Cl]_n$ and gives highly variable yields. The present procedure also gives $Pd_2(Ph_2PCH_2PPh_2)_2Cl_2$ free of contamination by $Pd_2(Ph_2PCH_2PPh_2)_2(\mu\text{-}CO)Cl_2$.[3]

*Department of Chemistry, University of California, Davis, CA 95616.
†Department of Chemistry, Case Western Reserve University, Cleveland, OH 44106.

A. DICHLOROBIS-μ-[METHYLENEBIS(DIPHENYL-PHOSPHINE)]-DIPALLADIUM(I)(Pd-Pd)

$$4Ph_2PCH_2PPh_2 + 2(PhCN)_2PdCl_2 + Pd_2(PhHC=CHC(O)CH=CHPh)_3 \cdot CHCl_3$$

$$\longrightarrow 2Pd_2(Ph_2PCH_2PPh_2)_2Cl_2 + 4PhCN + 3PhCH=CHC(O)CH=CHPh +$$

$$CHCl_3$$

Procedure

Under a nitrogen atmosphere, a 100-mL Schlenk flask containing a magnetic stirring bar is charged with 50 mL oxygen-free dichloromethane, bis(benzo-nitrile)dichloropalladium(II)[4] (0.412 g, 1.08 mmole), tris(1,5-diphenyl-1,4-pentadien-3-one)dipalladium chloroform solvate[5] (0.545 g, 0.527 mmole), and bis(diphenylphosphino)methane (0.819 g, 2.13 mmole). A reflux condenser is fitted to the flask, and the stirred solution is heated under reflux in a nitrogen atmosphere for 30 minutes. After cooling, the red solution, which at this stage is no longer oxygen sensitive, is filtered. The filtrate is condensed to a volume of 10 mL by the use of a rotary evaporator. Methanol (100 mL) is added to the dichloromethane solution to precipitate the product. The red-brown solid is collected by filtration and washed with methanol (3 × 10 mL) and diethyl ether (10 mL). The product is purified by dissolution in dichloromethane, filtration of the solution, and reprecipitation of the product by the addition of methanol. The orange-red, crystalline product is dried under vacuum.

Anal. Calcd. for $C_{25}H_{22}ClP_2Pd$: C, 57.06; H, 4.21; Cl, 6.74. Found: C, 57.40; H, 4.01; Cl, 7.00.

Properties

The compound $Pd_2(Ph_2PCH_2PPh_2)_2Cl_2$ is an orange-red solid which is moderately soluble in dichloromethane and chloroform and slightly soluble in benzene and toluene. It is air stable both as a solid and in solution. The ^1H nmr spectrum in CD_2Cl_2 solution exhibits the methylene resonance at 4.17 ppm. It is a 1:4:6:4:1 quintet (J_{P-H} = 4.0 Hz) due to virtual coupling of the protons to the four phosphorus atoms. The ^{31}P nmr spectrum in $CDCl_3$ solution consists of a singlet at -2.5 ppm with respect to external 85% H_3PO_4.[3] In solution, $Pd_2(Ph_2PCH_2PPh_2)_2Cl_2$ undergoes addition of a number of small molecules, including carbon monoxide,[3] isocyanides,[3] sulfur dioxide,[6] atomic sulfur,[6] and activated acetylenes.[7] Addition of these small molecules involves their insertion into the Pd—Pd bond. As a consequence, the Pd—Pd bond breaks and the Pd—Pd separation increases by ~0.5 Å. The following preparation offers an example of this behavior.

B. μ-CARBONYL-DICHLOROBIS[METHYLENEBIS(DIPHENYL-PHOSPHINE)] DIPALLADIUM(I)

$$Pd_2(Ph_2PCH_2PPh_2)_2Cl_2 + CO \longrightarrow Pd_2(Ph_2PCH_2PPh_2)_2(\mu\text{-}CO)Cl_2$$

Procedure

■ **Caution**: *Because of the very toxic nature of carbon monoxide, the following preparation should be conducted in an efficient fume hood.*

Dichlorobis[methylenebis(diphenylphosphine)] dipalladium(I) (0.10 g, 0.10 mmole) is dissolved in 10 mL dichloromethane. After filtration, carbon monoxide is slowly bubbled through the filtrate for 30 minutes. The product spontaneously precipitates as red-orange crystals. These are collected by filtration, washed with cold dichloromethane (5 mL) and diethyl ether (10 mL), and air dried. Yield 90%; mp 160° (dec.)

Anal. Calcd. for $C_{51}H_{44}Cl_2OP_4Pd_2$: C, 56.59; H, 4.10. Found: C, 56.59; H, 4.30.

Properties

The product forms moderately air-stable crystals which are slightly soluble in dichloromethane and chloroform. The infrared spectrum shows the bridging carbonyl absorption at 1705 cm^{-1} and the palladium-chloride stretching vibration at 258 cm^{-1}.[3] The solid is decarbonylated by heating at 78° under vacuum to re-form $Pd_2(Ph_2PCH_2PPh_2)_2Cl_2$. Decarbonylation may be effected from dichloromethane solutions by purging with a nitrogen stream or by heating under reflux.

C. TETRAKIS(1-ISOCYANOBUTANE)BIS[METHYLENEBIS(DIPHENYL-PHOSPHINE)] DIRHODIUM(I) BIS[TETRAPHENYLBORATE(1-)]

$$[(1,5\text{-}C_8H_{12})RhCl]_2 + 8n\text{-}C_4H_9NC + 2Na[BPh_4]$$

$$\longrightarrow 2[Rh(n\text{-}C_4H_9NC)_4][BPh_4] + 2NaCl + 2(1,5\text{-}C_8H_{12})$$

$$2[Rh(n\text{-}C_4H_9NC)_4][BPh_4] + 2Ph_2PCH_2PPh_2$$

$$\longrightarrow [Rh_2(Ph_2PCH_2PPh_2)_2(n\text{-}C_4H_9NC)_4][BPh_4]_2 + 4n\text{-}C_4H_9NC$$

Procedure

■ **Caution.** *Because of the toxicity and bad odor of butylisocyanide, this preparation should be performed in an efficient hood.*

1-Isocyanobutane (0.5 mL) is added to a stirred suspension of 0.25 g (0.50 mmole) dichlorobis(1,5-cyclooctadiene)dirhodium(I) in 15 mL methanol. The red-orange solution is filtered to remove any unreacted starting material. A solution of 0.41 g (1.2 mmole) sodium tetraphenylborate in 5 mL methanol is added to precipitate the product as yellow needles. The product is collected by filtration and washed with methanol. Purification of $[Rh(n\text{-}C_4H_9NC)_4][BPh_4]$ is achieved by recrystallization from dichloromethane/ethanol, mp 117-118°. The purple coloration of solutions of this yellow solid is caused by self-association of the planar monomer to form rhodium-rhodium-bonded dinuclear cations.

Anal. Calcd. for $C_{44}H_{56}BN_4Rh$: C, 70.03; H, 7.48. Found: C, 69.69; H, 7.53.

A solution of 0.132 g (0.343 mmole) bis(diphenylphosphino)methane in 10 mL acetone is added to a solution of 0.247 g (0.327 mmole) $[Rh(n\text{-}C_4H_9NC)_4]$-$[BPh_4]$ in 10 mL acetone. The blue-violet solution is filtered, and 15 mL 1-propanol is added to the filtrate. The solvent is partially removed from the mixture through the use of a rotary evaporator. Evaporation is stopped when purple crystals of the product begin to form. The product is collected by filtration and purified by recrystallization from acetone/1-propanol. Yield 0.19 g (60%).

Anal. Calcd. for $C_{59}H_{60}BN_2P_2Rh$: C, 72.85; H, 6.22; N, 2.88. Found: C, 72.48; H, 6.87; N, 2.98.

Properties

The compound $[Rh_2(Ph_2PCH_2PPh_2)_2(n\text{-}C_4H_9NC)_4][BPh_4]_2$ dissolves in dichloromethane, acetone, acetonitrile, and dimethylformamide to form intensely violet solutions. The complex may be identified from the isocyanide stretching absorption in the infrared spectrum at 2160 cm^{-1} (acetone solution) and from the following electronic absorptions: 560 nm (ϵ = 16,500), 338 (sh) (7700), 320 (18,100).[8] This complex reacts with iodine to form the brown complex $[Rh_2(Ph_2PCH_2PPh_2)_2(n\text{-}C_4H_9NC)_4I_2][BPh_4]_2$ via trans-annular oxidative addition.[8] It also reversibly forms an adduct with carbon monoxide.[9] Both of these reactions proceed with the strengthening of the rhodium-rhodium bond.

References

1. M. M. Olmstead, C. H. Lindsay, L. S. Benner, and A. L. Balch, *J. Organometal. Chem.*, **179**, 289 (1979).
2. R. Colton, R. H. Farthing, and M. J. McCormick, *Aust. J. Chem.*, **26**, 2607 (1973).
3. L. S. Benner and A. L. Balch, *J. Am. Chem. Soc.*, **100**, 6099 (1978).
4. M. S. Kharasch, R. C. Seyler, and F. R. Mayo, *J. Am. Chem. Soc.*, **60**, 882 (1938).
5. T. Ukai, H. Kawazura, Y. Ishii, J. J. Bonnet, and J. A. Ibers, *J. Organometal. Chem.*, **65**, 253 (1974).
6. A. L. Balch, L. S. Benner, M. M. Olmstead, *Inorg. Chem.*, **18**, 2996 (1979).
7. A. L. Balch, C.-H. Lee, C. H. Lindsay, and M. M. Olmstead, *J. Organometal. Chem.*, **177**, C22 (1979).

8. A. L. Balch, *J. Am. Chem. Soc.*, **98**, 8049 (1976).
9. J. T. Mague and S. H. DeVries, *Inorg. Chem.*, **19**, 3743 (1980).

12. SYNTHESES OF DIMETHYLAMIDO COMPOUNDS CONTAINING METAL-TO-METAL TRIPLE BONDS BETWEEN MOLYBDENUM AND TUNGSTEN ATOMS

Submitted by MALCOLM H. CHISHOLM,* DEBORAH A. HAITKO,* and
CARLOS A. MURILLO*
Checked by F. A. COTTON† and SCOTT HAN†

Hexakis(dimethylamido)dimolybdenum and -ditungsten compounds ($M \equiv M$) have proved to be important starting materials for entry into the rapidly developing area of the dinuclear chemistry of these Group VI transition elements.[1,2] Scheme 1 summarizes some of their reactions. We report here the details of our standard procedures for the preparation of two key compounds, $Mo_2(NMe_2)_6$ and $Mo_2Cl_2(NMe_2)_4$.[3,4] The tungsten analogs can be made in a similar manner.[4,5]

General Procedures and Techniques

The majority of starting materials and products are moisture and oxygen sensitive. Therefore, an inert atmosphere of dry and oxygen-free nitrogen was maintained throughout all experimental procedures. Solvents, except chlorobenzene, were distilled before use from either Na-K alloy or sodium benzophenone ketyl solutions. Chlorobenzene was dried and stored over 3-Å molecular sieves. Solvents were transferred by syringe, and bench top manipulations were performed using standard Schlenk techniques.[6] After preparation, the (dimethylamido)molybdenum compounds were handled in a Vacuum Atmospheres Company Drilab apparatus and stored in sealed glass ampuls. Where necessary, or appropriate, gases were measured on a calibrated vacuum manifold. Sublimations were performed using standard ground-glass apparatus: typically, a 50-mL, single-neck, round-bottomed flask was fitted to a 30-cm long glass tube with a gas inlet adapter. Reactive residues remaining after sublimation were destroyed by the careful addition of 2-propanol-petroleum ether solutions under an inert atmosphere.

*Department of Chemistry, Indiana University Bloomington, IN 47405.
†Department of Chemistry, Texas A&M University, College Station, TX 77843.

52

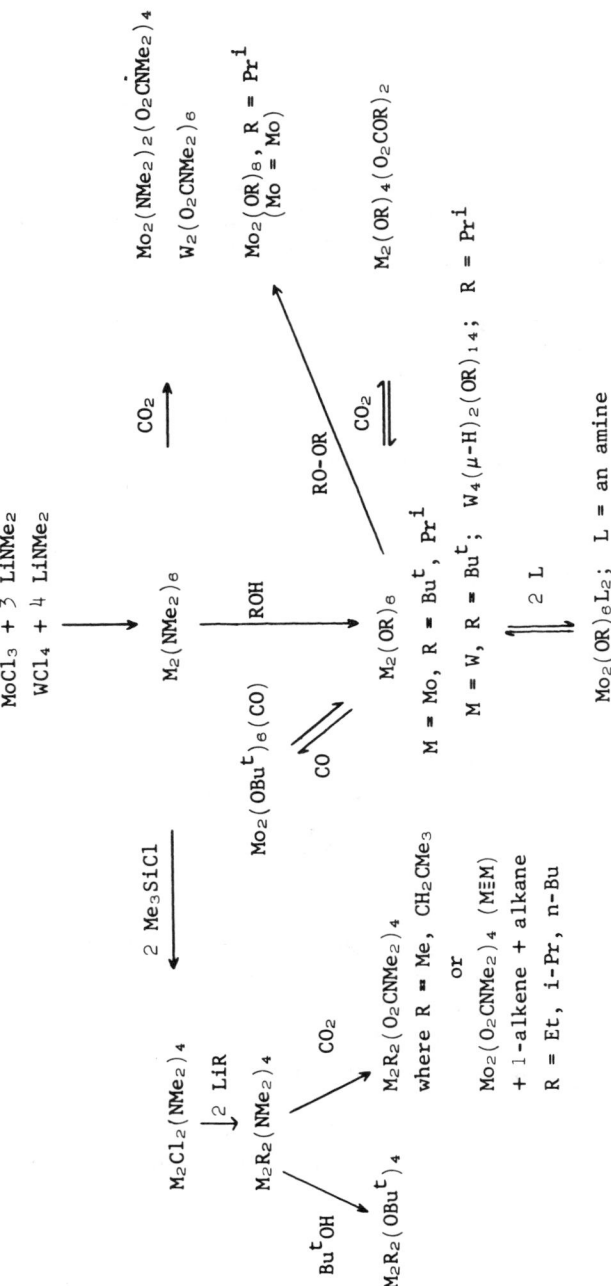

Scheme 1. Some reactions leading to and interconverting dinuclear compounds containing M—M triple bonds (M = Mo and W).

Starting Materials

$MoCl_3$ was prepared using the method of Mallock[7] from $MoCl_5$ and $SnCl_2$. These halides were purchased from ROC/RIC Co. An alternative route to $MoCl_3$ which we have developed follows the stoichiometry of the reaction

$$3MoCl_5 + 2Mo(CO)_6 \xrightarrow[\text{reflux}]{\text{PhCl}} 5MoCl_3 + 12CO\uparrow$$

The $MoCl_3$ prepared in this manner appears more reactive and gives greater yields of $Mo_2(NMe_2)_6$ than that prepared by $SnCl_2$ reduction of $MoCl_5$. (This observation was confirmed by the checkers.)

Chlorotrimethylsilane was purchased from Aldrich Chemical Co. and was distilled onto dried 3-Å molecular sieves prior to use.

Lithium dimethylamide was prepared from the reaction between butyllithium in hexane (commercially available from Alfa Ventron Co. as ca. 2.6 M LiBu in hexane) and dimethylamine according to the stoichiometric reaction:

$$\text{LiBu} + HNMe_2 \longrightarrow LiNMe_2 + n\text{-BuH}$$

The following is a typical preparation.

Butyllithium (210 mmole) in hexane (91 mL) was carefully transferred to a dried 500-mL, one-neck, round-bottomed flask with the use of a 50-mL-capacity syringe. The flask was equipped with a large Teflon-coated spinbar and was kept under a nitrogen atmosphere. The butyllithium solution was then cooled and frozen by immersing the flask in liquid nitrogen. The flask was evacuated, and, with the use of a calibrated vacuum manifold, dimethylamine (220 mmole) was added. When the addition of dimethylamine was complete, the flask was placed in a large Dry ice/2-propanol bath. The reaction vessel was kept at $-78°$ until the hexane solution no longer bubbled and a thick white precipitate was present. The reaction flask was then allowed to warm slowly to room temperature, at which point nitrogen was introduced into the system.

Alternate procedures, which involve either discharging anhydrous $HNMe_2$ from a cylinder directly into a solution of butyllithium until there is no further reaction or by the direct addition of a measured volume of liquid $HNMe_2$ cooled below its normal boiling point, $6°$, work equally well.

The lithium dimethylamide can be used without further purification in this hexane solution, or it may be stored as a fine, powdery solid by merely stripping off the hexane in vacuo.

■ **Caution.** *The hexane solution is saturated with butane, and the slurry of $LiNMe_2$ in hexane formed above should be handled with care while warming to room temperature.*

A. HEXAKIS(DIMETHYLAMIDO)DIMOLYBDENUM
(Mo≡Mo), Mo$_2$(NMe$_2$)$_6$

$$2MoCl_3 + 6LiNMe_2 \longrightarrow Mo_2(NMe_2)_6 + 6LiCl$$

To freshly prepared LiNMe$_2$ (hexane has been removed) (0.522 mole) in a 500-mL, round-bottomed flask equipped with a large Teflon-coated spinbar, tetrahydrofuran (200 mL) was added to give a pale-yellow solution. The solution was cooled to ca. 0° with the aid of an ice/water bath. The flask was fitted with a three-way adapter, a condenser (water-cooled), and a side-arm solids-addition tube. The configuration is shown in Fig. 1. MoCl$_3$ (37.2 g, 0.184 mole) was added slowly by way of the side-arm solids-addition tube; the solution was stirred magnetically. Upon addition of MoCl$_3$, the solution turned initially yellow brown and then dark brown. After 1 hour, the addition of MoCl$_3$ was complete and the solution was allowed to warm to room temperature slowly and left to stir for 12 hours (overnight). The solvent was stripped and the brown solids were thoroughly dried at 60°, 10^{-4} torr for 4 hours. Small quantities of the purple Mo(NMe$_2$)$_4$ sublimed out of the heated flask. The flask was then

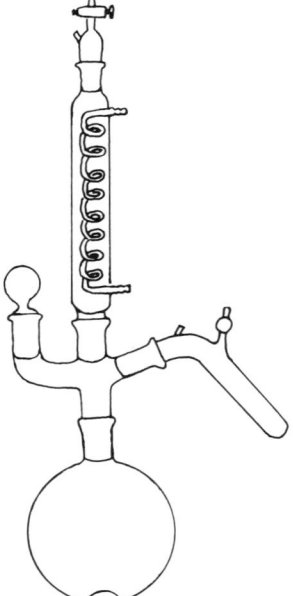

Fig. 1. *Reaction vessel.*

transferred to the dry box and the brown solids were scraped loose.

At this point, crude $Mo_2(NMe_2)_6$ can be obtained by hexane or pentane extraction of the solids using standard Schlenk filtration techniques. However, we have found that the following procedure has advantages in extracting all the $Mo_2(NMe_2)_6$, especially when very large-scale preparations are being carried out.

The solids were placed on the frit of a specially designed apparatus shown in Fig. 2. The extraction apparatus was assembled (Fig. 2) and hexane (200 mL) was placed in a 250-mL, round-bottomed flask equipped with a magnetic spin-bar. The nitrogen inlet adapter shown in Figure 2 was replaced by a short water-cooled condenser. The flask was then heated in an oil bath and hexane was dis-tilled onto the brown solids. In this manner, the $Mo_2(NMe_2)_6$ was removed from the crude product containing LiCl and other insoluble molybdenum-containing products. When the hexane returning to the round-bottomed flask was virtually colorless, the extraction was complete (2 hours). The hexane was stripped, and the 250-mL, round-bottomed flask was fitted with a tube sublimer and placed on a high-vacuum line. The solids were heated to 120° in an oil bath and over a period of 12 hours at 10^{-4} torr yellow crystalline $Mo_2(NMe_2)_6$ was collected in the sublimer. This was then scraped out and sealed in ampuls using a dry box facility. Yield 21.09 g, 50% based on the reaction stoichiometry.

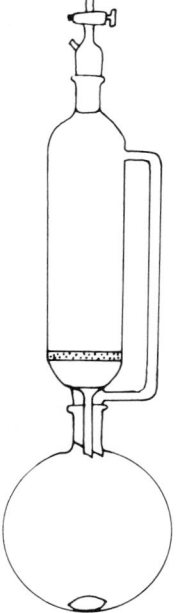

Fig. 2. Extraction apparatus.

Anal. Calcd. for $Mo_2(NMe_2)_6$: C, 31.59; H, 7.95; N, 18.42. Found: C, 31.29; H, 7.81; N, 18.17.

An alternate procedure for sublimation involves the use of a water-cooled conical sublimer. (This was recommended by the checkers.) The yellow, crystalline $Mo_2(NMe_2)_6$ can then be scraped from the walls.

Properties

Hexakis(dimethylamido)dimolybdenum(Mo≡Mo) is an air- and moisture-sensitive, yellow solid. It is appreciably soluble in hydrocarbon solvents: ca. 3 g per 100 mL hexane at 25°. The ^1H nmr spectrum is temperature dependent and reveals the large diamagnetic anisotropy associated with the molybdenum-to-molybdenum triple bond. At −50°, 60 MHz, in toluene-d_8, there are two signals of equal integral intensity at 4.13 and 2.41 ppm from HMDS, corresponding to proximal and distal methyl groups, respectively. (The $Mo_2(NC_2)_6$ unit has D_{3d} symmetry with the NC_2 blades aligned along the Mo-Mo axis.) Upon raising the temperature, these two signals initially broaden, then disappear in the baseline, and finally coalesce to a sharp singlet at δ = 3.25 ppm (rel. HMDS) above +50° as rotation about Mo−N bonds becomes rapid on the nmr time scale. Other spectroscopic data and physicochemical properties are reported in the literature.[3]

B.   DICHLOROTETRAKIS(DIMETHYLAMIDO)DIMOLYBDENUM (Mo≡Mo), $Mo_2Cl_2(NMe_2)_4$

$$Mo_2(NMe_2)_6 + 2Me_3SiCl \longrightarrow Mo_2Cl_2(NMe_2)_4 + 2Me_3SiNMe_2$$

Hexakis(dimethylamido)dimolybdenum(Mo≡Mo) (2.20 g, 4.8 mmole) was placed in a 100-mL, round-bottomed flask equipped with a nitrogen inlet and a magnetic Teflon-coated spinbar. A minimal amount of toluene was added, sufficient to dissolve all the $Mo_2(NMe_2)_6$ at room temperature (approx. 25 mL). The solution was cooled and frozen at liquid nitrogen temperature and the flask was placed under vacuum and attached to a calibrated vacuum manifold. Freshly distilled chlorotrimethylsilane (11.1 mmole) was added by slow condensation from the vacuum manifold. The reaction vessel was closed and left to warm slowly to room temperature. The solution was then stirred for 12 hours at 35°. With time, the color of the solution changed from pale yellow to orange and a finely divided microcrystalline precipitate of $Mo_2Cl_2(NMe_2)_4$ formed. The flask was then cooled to room temperature and placed in the refrigerator at −20°, yielding $Mo_2Cl_2(NMe_2)_4$ as a golden-yellow, finely divided crystalline solid which was collected by filtration (1.6 g, 70% yield). (The filtrate still contains

$Mo_2Cl_2(NMe_2)_4$ which can be collected by either further crystallization or by stripping the solvent and subliming the solid at $110°$, 10^{-4} torr.)

Anal. Calcd. for $Mo_2Cl_2(NMe_2)_4$: C, 21.88; H, 5.51; N, 12.76; Cl, 16.15. Found: C, 22.05; H, 5.43; N, 12.64; Cl, 16.07.

Properties

Dichlorotetrakis(dimethylamido)dimolybdenum($Mo\equiv Mo$) is a yellow, crystalline solid that is *rapidly* decomposed in the presence of oxygen and/or moisture. It has the anti-1,2-conformation, both in solution and in the solid state. The ^1H nmr spectrum is temperature dependent. At $-60°$ (220 MHz), two singlets of equal intensity are seen corresponding to proximal (δ = 4.21) and distal (δ = 2.39 ppm, relative to Me_4Si) methyl groups. Other chemical and physicochemical properties are given in the literature.[4]

References

1. F. A. Cotton and M. H. Chisholm, *Acc. Chem. Res.*, **11**, 356 (1978).
2. M. H. Chisholm, *Transition Metal Chem.*, **3**, 321 (1978).
3. M. H. Chisholm, F. A. Cotton, B. A. Frenz, W. W. Reichert, L. W. Shive, and B. R. Stults, *J. Am. Chem. Soc.*, **98**, 4469 (1976).
4. M. Akiyama, M. H. Chisholm, F. A. Cotton, M. W. Extine, and C. A. Murillo, *Inorg. Chem.*, **16**, 2407 (1977).
5. M. H. Chisholm, F. A. Cotton, M. W. Extine, and B. R. Stults, *J. Am. Chem. Soc.*, **98**, 4477 (1978).
6. D. F. Shriver, *The Manipulation of Air-Sensitive Compounds*, McGraw-Hill, New York, 1969.
7. A. K. Mallock, *Inorg. Synth.*, **12**, 178 (1970).

13. TETRANUCLEAR MIXED-METAL CLUSTERS

Submitted by G. L. GEOFFROY,* J. R. FOX,* E. BURKHARDT,* H. C. FOLEY,* A. D. HARLEY,* and R. ROSEN*
Checked by D. E. FJARE,† F. R. FURUYA,† J. W. HULL, JR.,† R. STEVENS,† and W. L. GLADFELTER†

Although the majority of cluster compounds have been prepared by routes in which chance has had a major role to play, there do exist several synthetic

*Department of Chemistry, Pennsylvania State University, University Park, PA 16802.
†Department of Chemistry, University of Minnesota, Minneapolis, MN 55455.

methods which allow the directed synthesis of particular clusters.[1-3] One of these is the redox condensation between a carbonylmetalate and a closed metal carbonyl trimer,[1,4-6] perhaps best exemplified by the high-yield (>95%) preparation of $[CoRu_3(CO)_{13}]^-$, Eq. 1[5]:

$$[Co(CO)_4]^- + Ru_3(CO)_{12} \longrightarrow [CoRu_3(CO)_{13}]^- + 3CO \qquad (1)$$

Details of this and the analogous syntheses of $FeRu_3H_2(CO)_{13}$ and $[FeRu_3H(CO)_{13}]^-$ are described herein.

Another excellent method involves the addition of coordinatively unsaturated monomeric complexes to the reactive Os=Os double bond of $Os_3H_2(CO)_{10}$, as illustrated below for the synthesis of $PtOs_3H_2(CO)_{11}(PPh_3)_2$, Eq. 2[7]:

$$Pt(C_2H_4)(PPh_3)_2 + Os_3H_2(CO)_{10} \longrightarrow PtOs_3H_2(CO)_{11}(PPh_3)_2 \qquad (2)$$

An adaptation of this method employing the addition of photogenerated $Fe(CO)_4$ and "$Ru(CO)_4$" to $Os_3H_2(CO)_{10}$ leads to $FeOs_3H_2(CO)_{13}$ and $RuOs_3H_2(CO)_{13}$, respectively, in good yield, and details of these syntheses are presented herein.

A. TRIDECACARBONYLDIHYDRIDOIRONTRIRUTHENIUM, $FeRu_3H_2(CO)_{13}$

$$Na_2[Fe(CO)_4] + Ru_3(CO)_{12} \longrightarrow Na_2[FeRu_3(CO)_{13}] + 3CO \qquad (3)$$

$$Na_2[FeRu_3(CO)_{13}] + H_3PO_4 \longrightarrow FeRu_3H_2(CO)_{13} + Na_2HPO_4 \qquad (4)$$

Solutions of the reactants in Eq. 3 are prepared in a N_2-filled glove box using tetrahydrofuran (THF) that has been distilled from Na benzophenone ketyl under N_2. $Ru_3(CO)_{12}$[8] (1.5 g, 2.33 mmole) and $Na_2[Fe(CO)_4] \cdot 1.5$ (1,4-dioxane) (Alfa-Ventron Corp.) (1.2 g, 3.47 mmole) are placed in separate 250-mL pressure-equalizing dropping funnels, and THF (200 mL) is added to each. A small magnetic stirbar is placed in each dropping funnel to aid in dissolution of the solid and to prevent clogging. The $Ru_3(CO)_{12}$ dropping funnel is capped with a gas inlet adapter, and the $Na_2[Fe(CO)_4]$ dropping funnel is capped with a glass stopper. All stopcocks are closed and the dropping funnels removed from the dry box. Alternatively, the solutions of reactants in the dropping funnels may be prepared using standard Schlenk techniques.

The reaction is conducted in a 1000-mL three-neck, round-bottom flask equipped with a magnetic stirbar, the two dropping funnels, and a reflux condenser capped with a gas inlet adapter. The three-neck flask is evacuated through the condenser and purged with N_2 3-4 times. The reaction is carried out under a N_2 atmosphere up to the point of the first extraction with hexane after acidification. The solution of $Na_2[Fe(CO)_4]$ is introduced into the three-neck flask and brought to reflux. The $Ru_3(CO)_{12}$ solution is then added dropwise over a 15 min period, during which time the initially tan reaction mixture turns dark red. After all the $Ru_3(CO)_{12}$ solution has been added, heating is continued for another 1.5 h.

■ **Caution.** *Carbon monoxide is evolved during this reaction and should be properly vented.*

The solvent is immediately removed under vacuum to leave a dark red-brown residual solid. Hexane (200 mL), deoxygenated by a N_2 purge, is added, followed by addition of deoxygenated 20% H_3PO_4 (100 mL). The hexane layer remains colorless until addition of acid, after which it becomes dark brown. Extraction with 200-mL portions of hexane is continued until the hexane layer is almost colorless. The hexane extracts are combined and dried over anhydrous $MgSO_4$, and the mixture is filtered and concentrated for chromatography.

Chromatographic separation of the reaction mixture on silica gel can be achieved in several ways. Rapid efficient separation is obtained in less than 1 hour using a Waters Associates model 500 preparative-scale high-pressure liquid chromatograph with their Prep PAK 500 Silica cartridges. The recommended flow rate is 200 mL hexane per minute for a 200-mL injection of the concentrated reaction mixture. Under these conditions, four fractions are generally obtained:

1. A yellow fraction containing both $Ru_3(CO)_{12}$ and $Ru_4H_4(CO)_{12}$ which elutes after ~10 min.
2. A green fraction of $Fe_3(CO)_{12}$ with a retention time of ~14 min.
3. A pink fraction of $Ru_4H_2(CO)_{13}$ which elutes after ~21 min.
4. A red fraction containing $FeRu_3H_2(CO)_{13}$ with a retention time of ~33 min.

Trace amounts of $FeRu_2(CO)_{12}$, $Fe_2Ru(CO)_{12}$, and $Fe_2Ru_2H_2(CO)_{13}$ may also elute before the $FeRu_3H_2(CO)_{13}$ fraction. If the $FeRu_3H_2(CO)_{13}$ fraction shows contamination with $Fe_2Ru_2H_2(CO)_{13}$, evidenced by an IR band at 2058 cm^{-1} the latter may be removed by recycling this fraction through the chromatograph for 2-3 cycles. If the first ~10% of the fraction is removed during each cycle, a very pure material may be obtained. High purity can also be achieved by stirring the $FeRu_3H_2(CO)_{13}$ fraction over silica gel in air for 1 day, during which time the $Fe_2Ru_2H_2(CO)_{13}$ impurity decomposes. The decomposition products of the compound apparently adhere to the silica gel.

Similar chromatographic separation of 100-200 mg of product can be achieved in ~1 day using an atmospheric pressure glass column (2 × 50 cm) with silica gel packing (Davison chromatographic grade H, 60-200 mesh), or more conveniently in several hours by using a 2.5 × 120 cm glass column (ACE 5820-40) packed with silica gel (Merck, 0.040-0.063 mm) and pressurized to 60 psig with a Fluid Metering Lab Pump model RP1SY/SS. Removal of the hexane solvent from the $FeRu_3H_2(CO)_{13}$ fraction gives the desired product in 56% yield (1.74 g). Contamination with $Fe_2Ru_2H_2(CO)_{13}$ is generally not a problem when the chromatographic separation is carried out using these latter methods, since $Fe_2Ru_2H_2(CO)_{13}$ decomposes on the silica gel during the time required for the separation.

Anal. Calcd. for $C_{13}H_2O_{13}FeRu_3$: C, 21.6; H, 0.28; Fe, 7.7. Found: C, 22.3; H, 0.43; Fe, 7.5.

Properties of $FeRu_3H_2(CO)_{13}$

$FeRu_3H_2(CO)_{13}$ is an air-stable, red, crystalline solid. It is soluble in nonpolar organic solvents such as hexane and very soluble in more polar organic solvents such as benzene and dichloromethane. Solutions of $FeRu_3H_2(CO)_{13}$ in these solvents are stable in air for several weeks. The compound is best characterized by its IR spectrum, which shows the following bands (cm^{-1}) in hexane solution: 2084 (s), 2072 (s), 2062 (w), 2040 (vs), 2030 (m), 2020 (w), 1991 (m).[4] A weak band at 2058 cm^{-1} may also appear, but this band is actually due to $Fe_2Ru_2H_2(CO)_{13}$. Thus, the relative intensity of this band is an indicator of the purity of $FeRu_3H_2(CO)_{13}$ since it should be absent in the spectrum of the pure material. The mass spectrum of $H_2FeRu_3(CO)_{13}$ shows a parent ion at m/e = 728 and ions corresponding to successive loss of all 13 carbonyls. UV-visible, ^{13}C, and ^1H nmr data have also been reported for $FeRu_3H_2(CO)_{13}$[1,9] as well as its crystal structure as determined by X-ray diffraction.[10]

B. μ-NITRIDO-BIS(TRIPHENYLPHOSPHORUS)(1+) TRIDECACARBONYL-HYDRIDOIRONTRIRUTHENATE(1−), $[\{(C_6H_5)_3P\}_2N][FeRu_3H(CO)_{13}]$

$$[\{(C_6H_5)_3P\}_2N][HFe(CO)_4] + Ru_3(CO)_{12} \longrightarrow$$

$$3CO + [\{(C_6H_5)_3P\}_2N][FeRu_3H(CO)_{13}]$$

Solid $Ru_3(CO)_{12}$[8] (0.277 g, 0.433 mmole), [PPN] $[HFe(CO)_4]$[11]([PPN]$^+$ = $[(Ph_3P)_2N]^+$) (0.306 g, 0.433 mmole), and a magnetic stirbar are added to a 250-mL, three-neck, round-bottom flask. A gas inlet adapter, reflux condenser topped with a gas inlet adapter, and a serum cap are then placed on the round-bottom flask, and the entire apparatus is evacuated and back-filled with nitrogen twice. From this point, all manipulations are carried out under a N_2 atmosphere. Tetrahydrofuran (THF), 100 mL, dried over Na benzophenone ketyl under N_2, is added to the round-bottom flask via a transfer needle. The solution is refluxed for 4 hours under a slow N_2 purge through the gas inlet adapter. A color change from orange to brown occurs during the reaction.

■ **Caution.** *Carbon monoxide is evolved during this reaction and should be properly vented.*

The THF is removed under vacuum, and 20 mL diethyl ether, dried over Na benzophenone ketyl under N_2, is added to dissolve the residue. This solution is filtered under N_2 into one side of a double-tube recrystallizer (Ace Glass Inc., No. 7772). Hexane (50 mL), dried over CaH_2 under N_2, is transferred via transfer needle to the other side of the double-tube recrystallizer through a rubber serum cap. This hexane solution (20 mL) is *slowly* layered over the Et_2O layer at $0°$. This is accomplished by pulling a slight vacuum on the Et_2O side of the double tube, then gently warming the hexane side until hexane begins to distill onto the Et_2O layer. When the hexane distilled over is of equal volume to the Et_2O, the layering procedure is stopped, and the layers are allowed to diffuse together overnight at $0°$. Black crystals of [PPN] [FeRu$_3$H(CO)$_{13}$] form at the interface of the ether/hexane layer. After opening the Et_2O/hexane side of the double-tube recrystallizer to the atmosphere, the solvent mixture is carefully decanted into a beaker, and the crystals of [PPN] [FeRu$_3$H(CO)$_{13}$] are carefully collected and washed three times with hexane and dried in vacuo to give 0.259 g (47% yield) of pure [PPN] [FeRu$_3$H(CO)$_{13}$].

Anal. Calcd. for $C_{49}H_{31}O_{13}NP_2FeRu_3$: C, 46.60; H, 2.48. Found: C, 45.94; H, 2.57.

Properties of [PPN][FeRu$_3$H(CO)$_{13}$]

[PPN] [FeRu$_3$H(CO)$_{13}$] is an air-stable, black, crystalline solid. It is insoluble in nonpolar organic solvents such as hexane and soluble in polar organic solvents such as diethyl ether, dichloromethane, and tetrahydrofuran. Solutions of [PPN] [FeRu$_3$H(CO)$_{13}$] are stable in air for several weeks. The compound is best characterized by its IR spectrum, which shows the following bands in dichloromethane solution: 2073 (w), 2032 (s), 2013 (s), 1998 (s), 1970 (sh), 1940 (sh), 1844 (w), 1809 (br) cm^{-1}. The ^{1}H nmr spectrum at $-80°$ in acetone-d_6 exhibits a single resonance at $\delta = -15.6$ ppm.[12] The crystal structure of the compound as determined by a neutron diffraction study has also been reported.[12]

C. μ-NITRIDO-BIS(TRIPHENYLPHOSPHORUS)(1+) TRIDECACARBONYL-COBALTTRIRUTHENATE(1−), [{(C$_6$H$_5$)$_3$P}$_2$N] [CoRu$_3$(CO)$_{13}$]

$$[\{(C_6H_5)_3P\}_2N] [Co(CO)_4] + Ru_3(CO)_{12} \longrightarrow$$
$$[\{(C_6H_5)_3P\}_2N] [CoRu_3(CO)_{13}] + 3CO$$

Procedure

A very high yield of [PPN] [CoRu$_3$(CO)$_{13}$] ([PPN]$^+$ = [(Ph$_3$P)$_2$N]$^+$) may be obtained using the apparatus shown in Fig. 1. Ru$_3$(CO)$_{12}$[8] (0.162 g, 0.253 mmole) is added to the pressure-equalizing dropping funnel and [PPN]-[Co(CO)$_4$][5] (0.167 g, 0.238 mmole) is placed into the 100-mL Schlenk flask. Both reagents are moderately air stable and can be handled for short periods in air in the solid state. The apparatus shown in Fig. 1 is assembled, except that the gas buret is not attached. The system is placed under a N$_2$ atmosphere and 30 mL tetrahydrofuran (THF), distilled under N$_2$ from Na benzophenone ketyl, is added to each reagent via a stainless steel transfer needle. The dropping funnel is purged briefly with N$_2$ to agitate and deaerate the Ru$_3$(CO)$_{12}$ suspension. The magnetically stirred [PPN] [Co(CO)$_4$] solution is brought to reflux and the system closed from the Schlenk manifold. The system is then vented to atmospheric pressure via the reflux condenser septum through a transfer needle and

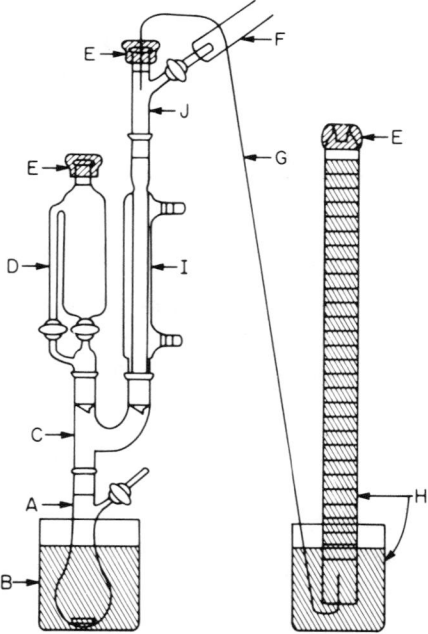

Fig. 1. Apparatus for the synthesis of [(Ph$_3$P)$_2$N][CoRu$_3$(CO)$_{13}$]. A) 100-mL Schlenk flask with magnetic stirbar; B) oil bath of dioctylphthalate in a 400-mL beaker; C) adapter; D) pressure equalizing funnel; E) septum cap; F) hose to Schlenk manifold; G) stainless steel transfer needle; H) 50-mL gas buret assembly filled with dioctylphthalate; I) Leibig condenser; and J) adapter with sidearm stopcock.

attached to the previously prepared 50-mL buret (oil-filled for displacement by slight evacuation through the serum cap), Fig. 1. Once the assembly is equilibrated, dropwise addition of the yellow-orange $Ru_3(CO)_{12}$ suspension produces a deep-red solution and an immediate evolution of CO gas. Addition is continued at a sufficient rate to maintain CO production until 16.0 mL is evolved (~0.5 hour). Reflux beyond this point will give a decreased yield. Although [PPN] [$CoRu_3(CO)_{13}$] is air stable in both the solid state and solution, exposure of the reactant solution to air during the preparation invariably gives poorer results.

■ **Caution.** *Carbon monoxide is evolved during this reaction and should be properly vented.*

The addition funnel and condenser assembly is replaced by a serum cap under a N_2 purge and the THF removed in vacuo at room temperature. The red, semicrystalline product is extracted into 30 mL diethyl ether and filtered. The compound may be recrystallized by slow evaporation in air or, preferably, by slow diffusion of petroleum ether into the ethereal solution using the double-tube recrystallization procedure described in the [PPN] [$FeRu_3H(CO)_{13}$] preparation. The former method gives a microcrystalline red-brown powder, whereas the latter, if conducted slowly, yields red,·prismatic crystals in near-quantitative yield (0.297 g; 0.235 mmole; 98.6% yield based on [PPN]-[$Co(CO)_4$]).

Anal. Calcd. for $C_{49}H_{30}O_{13}NP_2CoRu_3$: C, 46.53; H, 2.40; Ru, 23.97. Found: C, 46.55; H, 2.38; Ru, 23.65 (Alfred Bernhardt Analytical Laboratories, Engleskirchen, West Germany).

Properties of [PPN][CoRu₃(CO)₁₃]

[PPN] [$CoRu_3(CO)_{13}$] is an air-stable, red crystalline solid, insoluble in nonpolar organic solvents such as hexane but highly soluble in polar organic solvents such as CH_2Cl_2, Et_2O, THF, acetone, MeOH, and EtOH. Characterization is best accomplished by its IR spectrum: ν_{CO} (THF): 2067 (w), 2018 (vs), 1993 (m), 1974 (w, sh), 1828 (w), 1803 (w, br) cm^{-1}.[5] The crystal structure of [PPN] [$CoRu_3(CO)_{13}$] has also been reported.[5]

D. TRIDECACARBONYLDIHYDRIDOIRONTRIOSMIUM, $FeOs_3H_2(CO)_{13}$

$$Fe(CO)_5 + Os_3H_2(CO)_{10} \xrightarrow{h\nu} FeOs_3H_2(CO)_{13} + 2CO$$

Procedure

A solution (100 mL) of $Os_3H_2(CO)_{13}$ (77.2 mg, 0.090 mmole)[13] and $Fe(CO)_5$ (Alfa-Ventron Corp.) (100 μl, ~0.743 mmole) in hexane, previously dried and degassed by distillation from Na benzophenone ketyl, is prepared in a 250-mL, three-neck, round-bottom flask equipped with a magnetic stirbar, gas inlet adapter, reflux condenser capped with a gas inlet adapter, and a ground-glass stopper. The solution is stirred while irradiated with 366-nm light (Blak-Ray B-100 A Long Wave UV Lamp, Ultraviolet Products, San Gabriel, CA) for 2 hours under a slow N_2 purge through the gas inlet adapters. During this time, the solution changes color from purple to orange, and an orange film forms on the side of the flask after about 0.5 hours of irradiation.

■ **Caution.** *Carbon monoxide is evolved during this reaction and should be properly vented.*

The solvent and unreacted $Fe(CO)_5$ is removed under vacuum to leave an orange-brown residue. Chromatographic separation of the product can be achieved by dissolving the residue in 40 mL of a 9:1 hexane/benzene solvent mixture and eluting with the same solvent mixture on a 2 × 30 cm glass column (ACE 5820-16) packed with silica gel (Merck, 0.040-0.063 mm) and pressurized to 15 psig with a Fluid Metering Lab Pump model RP1SY/SS pump.

Under these conditions, three fractions are obtained in the following order:

1. A green fraction containing $Fe_3(CO)_{12}$.
2. A purple fraction containing unreacted $Os_3H_2(CO)_{10}$.
3. An orange fraction containing $FeOs_3H_2(CO)_{13}$.

Removal of the solvent from the $FeOs_3H_2(CO)_{13}$ fraction gives the desired product in 95% yield (85.4 mg). This material is sufficiently pure for most purposes, and further recrystallization is unnecessary.

Properties of $FeOs_3H_2(CO)_{13}$

$FeOs_3H_2(CO)_{13}$ is a moderately air-stable, orange, crystalline solid. It is somewhat soluble in nonpolar organic solvents such as hexane and very soluble in more polar solvents such as benzene and dichloromethane. The compound is best characterized by its IR spectrum, which shows the following bands in hexane solution: 2114 (vw), 2087 (s), 2073 (s), 2041 (vs), 2034 (m), 2027 (m), 2017 (m), 1994 (m), 1875 (w), 1846 (m) cm^{-1}.[14,15]

E. TRIDECACARBONYLDIHYDRIDORUTHENIUMTRIOSMIUM, $RuOs_3H_2(CO)_{13}$

$$Ru_3(CO)_{12} + 3Os_3H_2(CO)_{10} \xrightarrow{h\nu} 3RuOs_3H_2(CO)_{13} + 3CO$$

A hexane solution (80 mL) of $Os_3H_2(CO)_{10}$ (68.9 mg, 0.081 mmole)[13] and

$Ru_3(CO)_{12}$ (53.1 mg, 0.830 mmole) is prepared in a 100-mL, three-neck, round-bottom flask equipped in the same manner as in the preparation of $FeOs_3H_2(CO)_{13}$ described above. The solution is irradiated with 366-nm light, as in the previous $FeOs_3H_2(CO)_{13}$ preparation, for 70 hours. The solution is then reduced in volume to about 30 mL by evaporation of solvent under vacuum.

- **Caution.** *Carbon monoxide is evolved during this reaction and should be properly vented.*

Chromatography is carried out using the apparatus previously described in the $FeOs_3H_2(CO)_{13}$ preparation except that neat hexane is used as the eluant. Four fractions are obtained in the following order:

1. A yellow-brown inseparable mixture of $Ru_3(CO)_{12}$ and $Os_3H_2(CO)_{10}$.
2. A red-orange fraction containing $Ru_4H_2(CO)_{13}$.
3. An orange fraction of $RuOs_3H_2(CO)_{13}$.
4. A yellow fraction containing $Os_4H_4(CO)_{12}$.

Removal of the hexane solvent from the $RuOs_3H_2(CO)_{13}$ fraction gives the desired product in 53% yield (44.0 mg based on $Os_3H_2(CO)_{10}$).

Anal. Calcd. for $C_{13}H_2O_{13}RuOs_3$: C, 15.04%; H, 0.19%. Found: C, 15.04%; H, 0.15%.

Properties of $RuOs_3H_2(CO)_{13}$

$RuOs_3H_2(CO)_{13}$ is an air-stable, orange, crystalline solid. It is soluble in nonpolar organic solvents such as hexane as well as more polar organic solvents such as benzene and dichloromethane. Solutions of $RuOs_3H_2(CO)_{13}$ are air stable indefinitely. The compound is best characterized by its IR spectrum, which shows the following bands in hexane solution: 2110 (vw), 2082 (vs), 2067 (vs), 2057 (vs), 2029 (m), 2025 (m), 2019 (s), 2008 (w), 1869 (w, br) cm^{-1}. The ^1H nmr spectrum at 25° in acetone-d_6 solvent shows a broad singlet at δ -20.7 ppm (6.7 Hz half-width).[15]

References

1. W. L. Gladfelter and G. L. Geoffroy, *Adv. Organometal. Chem.*, **18**, 207 (1980).
2. P. Chini, G. Longoni, and V. G. Albano, *Adv. Organometal. Chem.*, **14**, 285 (1976).
3. P. Chini and B. T. Heaton, *Topics Curr. Chem.*, **71**, 1 (1977).
4. G. L. Geoffroy and W. L. Gladfelter, *J. Am. Chem. Soc.*, **99**, 7565 (1977).
5. P. C. Steinhardt, W. L. Gladfelter, A. D. Harley, J. R. Fox, and G. L. Geoffroy, *Inorg. Chem.*, **19**, 332 (1980).
6. J. Knight and M. J. Mays, *J. Chem. Soc., Dalton Trans.*, 1022 (1972).
7. L. J. Farrugia, J. A. K. Howard, P. Matuprochochon, J. L. Spencer, F. G. A. Stone and P. Woodward, *J. Chem. Soc., Chem. Commun.*, 260 (1978).
8. A. Mantovani and S. Cerrini, *Inorg. Synth.*, **13**, 93 (1972).

9. G. L. Geoffroy and W. L. Gladfelter, *Inorg. Chem.*, **19**, 2579 (1980).

10. C. J. Gilmore and P. Woodward, *J. Chem. Soc., A*, 3453 (1971).

11. M. Y. Darensbourg, D. J. Darensbourg and H. L. C. Barros, *Inorg. Chem.*, **17**, 297 (1978).

12. F. Takusagawa, A. Fumagalli, T. F. Koetzle, G. R. Steinmetz, R. P. Rosen, W. L. Gladfelter, G. L. Geoffroy, M. A. Bruck and R. Bau, *Inorg. Chem.*, **20**, 3823 (1981).

13. S. A. R. Knox, J. W. Koepke, M. A. Andrews and H. D. Kaesz, *J. Am. Chem. Soc.*, **97**, 3942 (1975).

14. J. R. Moss and W. A. G. Graham, *J. Organometal. Chem.*, **23**, C23 (1970).

15. E. W. Burkhardt and G. L. Geoffroy, *J. Organometal. Chem.*, **198**, 179 (1980).

14. BIS[μ-NITRIDO-BIS(TRIPHENYLPHOSPHORUS)(1+)] TRIDECACARBONYLTETRAFERRATE(2-) OR [BIS(TRIPHENYLPHOSPHINE)IMINIUM TRIDECACARBONYL-TETRAFERRATE] (2-) [PPN] $_2$ [Fe$_4$(CO)$_{13}$]

Submitted by K. WHITMIRE,* J. ROSS,* C. B. COOPER, III,* and D. F. SHRIVER*
Checked by JOHN S. BRADLEY† (Part A) and W. J. COTE‡ and P. J. KRUSIC‡ (Part B)

The tetranuclear dianion [Fe$_4$(CO)$_{13}$]$^{2-}$ was first isolated by Hieber and Werner[1] as the hexakis(pyridine)iron(II) salt. Because of the instability of the cation, several other salts were synthesized.[1] The general procedure described here is based on the original synthesis, but reaction conditions have been optimized to attain higher yield and shorten the reaction time (1 hour rather than 24 hours). In method A, the conversion of the hexakis(pyridine)iron(II) salt to the PPN salt has been greatly simplified and eliminates the use of hydroxide ion, which causes side reactions. Method B illustrates a procedure employing an improved precipitation of iron(II) hydroxide, which provides a general route to [Fe$_4$(CO)$_{13}$]$^{2-}$ salts with many different cations.

Procedure

$$5Fe(CO)_5 + 6C_5H_5N \longrightarrow [Fe(C_5H_5N)_6] [Fe_4(CO)_{13}] + 12CO$$

■ **Caution.** *The reaction must be run in a hood because of the toxicity of iron carbonyls and carbon monoxide.*

Fresh iron pentacarbonyl may be used without purification. Pyridine is

*Department of Chemistry, Northwestern University, Evanston, IL 60201.

†Exxon Research and Engineering Corp., Box 45, Linden, NJ 07036.

‡Central Research and Development Dept., E. I. du Pont de Nemours and Co., Wilmington, DE 19898.

purified by storing it over potassium hydroxide and then distilling it from barium oxide. Acetonitrile is purified by distillation from phosphorus pentoxide. Methanol is purified by fractional distillation, and anhydrous diethyl ether is prepared by distillation from Na benzophenone ketyl. All solvents are distilled under nitrogen and purged with nitrogen before use.

Bis(triphenylphosphine)iminium chloride [PPN]Cl, which is easily synthe-sized[3] or available commercially (Alfa Products, Danvers, Massachusetts; the systematic IUPAC name for this salt is μ-nitrido-bis(triphenylphosphorus)(1+), is purified by dissolution in hot water, addition of activated charcoal, filtration of the solution, and crystallization from the cooling solution. The resulting crystalline [PPN]Cl is filtered and dried under high vacuum overnight.

All of the following operations are performed in a nitrogen-flushed dry box or in standard Schlenk apparatus under pure, dry nitrogen atmosphere.[2]

A double-neck, round-bottom, 100-mL flask containing a magnetic stirring bar is fitted with a water-cooled reflux condenser. The condenser is topped with a T-tube which connects to a nitrogen source and a pressure-release oil bubbler. A 20-mL sample of $Fe(CO)_5$, which has been degassed by several cycles of freezing, pumping, and thawing, is introduced by syringe into the flask. A 35-mL sample of degassed pyridine is then added. The flask is then immersed in a hot oil bath (115-120°), and the solution is allowed to reflux for 1 hour. The color of the solution changes from yellow to dark red after approximately 10 minutes. After cooling, the product is filtered through a medium-porosity Schlenk-type frit and washed with 10-mL portions of diethyl ether until the filtrate is colorless. The product, hexakis(pyridine)iron(II) tridecacarbonyl-tetraferrate, is dried for 10 minutes under vacuum (because of the lability of the coordinated pyridine, longer evacuation causes decomposition). Yield 22.6 g (68%).

Anal. Calcd. for $C_{43}H_{30}N_6Fe_5O_{13}$: C, 46.19; H, 2.70; N, 7.52; Fe, 24.97. Found: C, 44.86; H, 2.77; N, 7.36; Fe, 26.77.

Method A

$$[Fe(C_5H_5N)_6] [Fe_4(CO)_{13}] + 2[PPN]Cl \xrightarrow{\ CH_3OH\ } [PPN]_2[Fe_4(CO)_{13}]$$

$$+ \text{other products}$$

To a 100-mL, round-bottom flask containing a magnetic stirrer, 30 mL CH_3OH and 3.0 g (2.7 mmole) $[Fe(py)_6][Fe_4(CO)_{13}]$ is added, and the solution is stirred while 3.3 g (5.7 mmole) [PPN]Cl dissolved in 10 mL methanol is added. The dark, red-black product is collected by filtration on a medium-porosity Schlenk-type fritted filter and then dried under vacuum. This product is then purified by dissolving it in a minimum of CH_3CN (ca. 25 mL), filtering through a medium-porosity Schlenk-type fritted filter, and then inducing it to crystallize by the addition of methanol (in excess of 8 times the volume of CH_3CN used). The product is collected as before and dried under vacuum for 15 minutes. Yield 2.80 g (62%).

Anal. Calcd. for $C_{85}H_{60}N_2Fe_4O_{13}P_4$: C, 61.32; H, 3.63; N, 1.68. Found: C, 60.80; H, 3.03; N, 1.87.

Method B

$$[Fe(C_5H_5N)_6][Fe_4(CO)_{13}] + 2KOH \longrightarrow K_2[Fe_4(CO)_{13}]$$

$$\xrightarrow{2[PPN]Cl} [PPN]_2[Fe_4(CO)_{13}] \downarrow$$

To a 100-mL, round-bottom flask containing a magnetic stirrer, 20 mL CH_3CN and 1 g (0.90 mmole) of the hexakis(pyridine)iron(II) salt, 5 mL of a 0.4 M methanolic KOH is added, and the solution is stirred for 15 minutes. Acetonitrile is employed to produce a more easily filtered precipitate. The solution is filtered through a medium-porosity Schlenk-type frit, and the residue is washed with two 5-mL portions of acetonitrile. (A crude form of $K_2[Fe_4(CO)_{13}]$ may be prepared by vacuum evaporation of this solution.) To the filtrate is added 1.2 g (2.1 mmole) [PPN]Cl dissolved in 5 mL acetonitrile, and the solution is filtered through a medium-porosity Schlenk-type frit. The precipitate is washed with 5 mL acetonitrile, and the filtrate plus washings is then concentrated under vacuum to a total volume of approximately 5 mL. To this solution is added 50 mL diethyl ether and the precipitate is collected by filtration. The product is recrystallized and dried as in method A. Yield 1.22 g (82%).

Anal. Calcd. for $C_{85}H_{60}N_2Fe_4O_{13}P_4$: C, 61.32; H, 3.63; N, 1.68; Fe, 13.42. Found: C, 59.36; H, 3.62; N, 1.59; Fe, 12.68.

A variety of tetraalkylammonium salts may be precipitated from an aqueous solution of the crude potassium salt, $K_2[Fe_4(CO)_{13}]$, by the addition of an aqueous solution of a tetraalkylammonium halide.

Properties

Bis(triphenylphosphine)iminium   tridecacarbonyltetraferrate(2-), $[PPN]_2$-$[Fe_4(CO)_{13}]$, is obtained as shiny, very dark, red-black crystals. The compound is moderately air sensitive in solid form and decomposes slowly in solution when exposed to air, complete decomposition requiring several hours. It is soluble in CH_2Cl_2, CH_3CN, and pyridine and is insoluble in diethyl ether, benzene, tetrahydrofuran, ethanol, and methanol. It can be stored for long periods of time in an inert atmosphere (e.g., N_2) at room temperature. The IR spectrum in dichloromethane solution contains the following CO stretching frequencies: 2020 (w), 1943 (s), 1778 (w), 1682 (w), cm^{-1} (w = weak, s = strong).

The structure of the $Fe_4(CO)_{13}{}^{2-}$ has been established by X-ray diffraction on the hexakis(pyridine) salt,[4] and the Mössbauer spectrum has been determined for the tetraethylammonium salt.[5] Acidification of the anion produces the monohydride anion, $[Fe_4(CO)_{13}H]^-$,[5] of known structure.[6] The dihydride has also been reported to result on further acidification.[1,7] The dianion reacts with CH_3SO_3F to yield $[PPN][Fe_4(CO)_{12}(\mu_3\text{-}COCH_3)]$,[8] which can be protonated giving $HFe_4(CO)_{12}(\eta^2\text{-}COCH_3)$.[9]

References

1. W. Hieber and R. Werner, *Chem. Ber.*, **90**, 286 (1957).
2. D. F. Shriver, *The Manipulation of Air-Sensitive Compounds*, McGraw-Hill Book Company, New York, 1969, Part 2.
3. R. K. Ruff and W. J. Schlientz, *Inorg. Synth.*, **15**, 84 (1974).
4. R. J. Doedens and L. F. Dahl, *J. Am. Chem. Soc.*, **88**, 4847 (1966).
5. K. Farmery, M. Kilner, R. Greatrex, and N. N. Greenwood, *J. Chem. Soc., A*, 2339 (1969).
6. M. Manassero, M. Sansoni, and G. Longoni, *J. Chem. Soc., Chem. Comm.*, 919 (1976).
7. K. H. Whitmire and D. F. Shriver, *J. Am. Chem. Soc.*, **103**, 6754 (1981).
8. E. M. Holt, K. Whitmire, and D. F. Shriver, *J. Chem. Soc., Chem. Comm.*, 778 (1980).
9. K. Whitmire, D. F. Shriver, and E. M. Holt, *J. Chem. Soc., Chem. Comm.*, 780 (1980).

Chapter Three

ORGANOMETALLIC COMPOUNDS

15. CHLOROBIS(PENTAFLUOROPHENYL)THALLIUM(III)

Submitted by RAFAEL USON* and ANTONIO LAGUNA*
Checked by JOHN L. SPENCER† and DAVID G. TURNER†

Bromobis(pentafluorophenyl)thallium(III) has been obtained in only moderate yields ($\sim$45%) by the reaction of $TlCl_3$ with C_6F_5MgBr, which leads to a quite dark product that must repeatedly be recrystallized, the overall procedure not taking less than 16 hours. The chlorobis(pentafluorophenyl)thallium derivative has been prepared[1] by reacting the bromo derivative with AgCl. Both thallium(III) complexes are capable of transferring their two C_6F_5 groups to complexes of transition or post-transition metals, and can therefore be employed as arylating agents in the preparation of novel pentafluorophenyl complexes[2-6] whose oxidation states are generally higher than those of the parent compounds. Herein we describe a straightforward method which in only ca. 8 hours allows the synthesis and isolation of $Tl(C_6F_5)_2Cl$ in higher yields (80-85%).

Procedure

■ **Caution.** *Since thallium salts are harmful and can be absorbed through the skin, the use of Neoprene gloves is recommended during the manipulation.*

*Department of Inorganic Chemistry, University of Zaragoza, Spain.
†Department of Inorganic Chemistry, University of Bristol, Bristol BS8 1TS, England.

The used solvents (diethyl ether and benzene) are previously dried over sodium wire and distilled under nitrogen.

A. PREPARATION OF ANHYDROUS THALLIUM(III) CHLORIDE

$$TlCl_3 \cdot 4H_2O + 4SOCl_2 \longrightarrow TlCl_3 + 4SO_2 + 8HCl$$

A round-bottom, two-neck, 100-mL flask, placed in an isopropyl alcohol-Dry ice bath ($-75°$), is fitted with a Teflon-coated magnetic stirring bar, reflux condenser cooled with isopropyl alcohol-Dry ice (250 mL), and a gas outlet connected to an oil bubbler. Dry Cl_2, 3 mL, introduced through the second neck, is condensed, and 20 g (52.2 mmole) commercial thallium trichloride ($TlCl_3 \cdot 4H_2O$) and an excess of sulfinyl chloride (30 mL) are successively added while the mixture is rapidly stirred, whereafter the neck is closed with a stopper. The stirred mixture is slowly warmed to room temperature (5 hours). The excess sulfinyl chloride is decanted and the remaining white powder is vacuum dried to give 15.8 g (98%) $TlCl_3$, which must be stored in vacuo over phosphorus pentoxide.

B. PREPARATION OF (PENTAFLUOROPHENYL)LITHIUM[7]

$$C_6F_5Br + Bu^nLi \longrightarrow LiC_6F_5 + Bu^nBr$$

A round-bottom, two-neck, 250-mL flask fitted with a Teflon-coated magnetic stirring bar, an inlet for dry nitrogen with a stopcock, and a pressure-equalizing dropping funnel connected to a mineral oil bubbler is cooled to $-75°$ (isopropyl alcohol-Dry ice) and charged with 7.0 mL (55 mmole) bromopenta-fluorobenzene in 150 mL diethyl ether, whereupon a solution of 55 mmole butyllithium[8] in hexane is added slowly (about 1 drop per second). As soon as the addition is completed, the dropping funnel is replaced by a mineral oil bubbler and the mixture is stirred in a nitrogen atmosphere for 1 hour while the temperature is maintained at $-75°$.

C. PREPARATION OF CHLOROBIS(PENTAFLUOROPHENYL)-THALLIUM(III)

$$TlCl_3 + 2LiC_6F_5 \longrightarrow Tl(C_6F_5)_2Cl + 2LiCl$$

Anhydrous thallium trichloride, 7.8 g (25 mmole), is added to the above solution of freshly prepared (pentafluorophenyl)lithium and the mixture is stirred for 30 minutes at $-75°$, after which it is allowed to slowly warm to room temperature (1 hour). The nitrogen stream is stopped and the precipitated lithium chloride is removed by filtration. The yellow-orange filtrate is treated with 80 mL water *previously* acidified with 4 drops of concentrated HCl (addition of water followed by acidification leads to OH-containing end products). The organic layer, which during this operation has lost color, is removed by using a separating funnel, and the aqueous layer is extracted with two 10-mL portions of diethyl ether, which are then added to the previously separated diethyl ether solution and dried with 1-2 g anhydrous magnesium sulfate. After filtration, the diethyl ether is evaporated, and the resulting white residue is extracted with four 10-mL portions of benzene at 80°. (If toluene instead of benzene is used in order to reduce the health risks, the preparation renders lower yields, 65-70%.) (■ **Caution.** *Benzene vapors are noxious, also possibly carcinogenic, and should be handled in a fume hood.*) The extract is filtered through a hot sintered-glass frit (porosity 2) into a round bottom 100 mL flask. On cooling a white solid crystallizes, which is filtered off, washed with three 10 mL portions of hexane and finally vacuum-dried (9.8 g). Evaporation of the mother-liquors to about half the original volume renders new amounts of the crystals. Combined yield is 11.9 g(83%).

Anal. Calcd. for $C_{12}ClF_{10}Tl$: C, 25.1. Found: C, 25.2

Properties

Chlorobis(pentafluorophenyl)thallium(III) is a white, crystalline solid which is indefinitely air and moisture stable, mp 237-239°. It is soluble in acetone, diethyl ether, dichloromethane, chloroform, and warm benzene; it is slightly soluble in cold benzene and insoluble in hexane. It is dimeric in chloroform solution (MW 1162, calcd. 574). Characteristic strong bands occur in the infrared spectrum at 1518, 1090, and 970 cm^{-1}. The alkaline hydrolysis of $TlCl(C_6F_5)_2$ gives $Tl(OH)(C_6F_5)_2$ [ν(OH) at 3560 cm^{-1}].

Arylating and oxidizing properties of $TlCl(C_6F_5)_2$ are similar to those of the bromo derivative. It is capable of transferring two C_6F_5 groups to an appropriate substrate, that is, to gold(I) or tin(II) complexes, forming complexes whose central atom is in an oxidation state which is 2 units higher than in the starting compound.[9]

Analogous Complexes

Other thallium compounds of the type TlR_2Cl (R = 2,3,4,6-C_6F_4H, 2,3,5,6-C_6F_4H, or 2,4,6-$C_6F_3H_2$) can also be prepared by this method. Since they are more soluble in cold benzene, it is convenient to recrystallize them from a mixture of dichloromethane-hexane. (The yields are ca. 67, 75, and 75%, respectively.)

References

1. G. B. Deacon, J. H. S. Green, and R. S. Nyholm, *J. Chem. Soc.*, 3411 (1965).
2. G. B. Deacon and J. C. Parrott, *J. Organometal. Chem.*, **17**, P17 (1969).
3. R. S. Nyholm and P. Royo, *J. Chem. Soc. Chem. Comm.*, 421 (1969).
4. R. Usón, P. Royo, and A. Laguna, *J. Organometal. Chem.*, **69**, 361 (1974).
5. R. Usón, P. Royo, J. Forniés, and F. Martínez, *J. Organometal. Chem.*, **90**, 367 (1975).
6. F. Caballero and P. Royo, *Synth. React. Inorg. Metalorg. Chem.*, **7**, 351 (1976).
7. P. L. Coe, R. Stephens, and J. C. Tatlow, *J. Chem. Soc., A*, 3227 (1962).
8. E. H. Amonoo-Neizer, R. A. Shaw, D. O. Skovlin, and B. C. Smith, *Inorg. Synth.*, 8, 19 (1966).
9. R. Usón, A. Laguna, and T. Cuenca, *J. Organometal. Chem.*, **194**, 271 (1980).

16. (η^6-HEXAMETHYLBENZENE)RUTHENIUM COMPLEXES

Submitted by M. A. BENNETT,* T.-N. HUANG,* T. W. MATHESON,* and A. K. SMITH†
Checked by STEVEN ITTEL‡ and WILLIAM NICKERSON‡

Areneruthenium(II) complexes, $[Ru(\eta^6\text{-arene})Cl_2]_2$,[1-3] have interesting catalytic properties[4-7] and are useful synthetic precursors to a range of arene complexes of zerovalent and divalent ruthenium.[2,3,6-8] Most of the complexes can be made by reaction of ethanolic ruthenium trichloride with the appropriate 1,3- or 1,4-cyclohexadiene.[1-4] However, the hexamethylbenzene complex, $\{RuCl_2[\eta^6\text{-}C_6(CH_3)_6]\}_2$, one of the more soluble and stable members of the series, cannot be prepared in this way because hexamethylbenzene cannot be reduced easily to the corresponding 1,4-diene by dissolving metal reduction. It is necessary first to prepare the 1-isopropyl-4-methylbenzene (*p*-cymene) complex, $\{RuCl_2[\eta^6\text{-}1\text{-}[CH(CH_3)_2]\text{-}4\text{-}CH_3C_6H_4]\}_2$, by reaction of ruthenium trichloride with commercially available α-phellandrene (5-isopropyl-2-methyl-1,3-cyclohexadiene), and the coordinated *p*-cymene can then be displaced quantitatively with hexamethylbenzene.

*Research School of Chemistry, The Australian National University, Canberra, A.C.T., Australia 2600.
†Department of Inorganic Chemistry, University of Liverpool, Liverpool, England L69 3BX.
‡Central Research and Development Dept., Experimental Station, E. I. du Pont de Nemours and Co., Wilmington, DE 19898.

A. DI-μ-CHLORO-BIS[CHLORO(η^6-1-ISOPROPYL-4-METHYL-BENZENE)RUTHENIUM(II)]

$$2RuCl_3 + 2C_{10}H_{16} \xrightarrow{C_2H_5OH} [RuCl_2(\eta^6\text{-}C_{10}H_{14})]_2 + \cdots$$

Procedure

A solution of hydrated ruthenium trichloride (approximating $RuCl_3 \cdot 3H_2O$, containing 38-39% Ru, available from Johnson-Matthey Co. Ltd., 78 Hatton Garden, London, England EC1N 8EE) (2 g, approx. 7.7 mmole) in 100 mL ethanol is treated with 10 mL α-phellandrene (Fluka AG, CH-9470, Buchs SG, Switzerland) and heated under reflux in a 150-mL, round-bottomed flask for 4 hours. A nitrogen atmosphere can be used but is not strictly necessary. The solution is allowed to cool to room temperature, and the red-brown, micro-crystalline product is filtered off. Additional product is obtained by evaporating the orange-yellow filtrate under reduced pressure to approximately half-volume and refrigerating overnight. After drying in vacuo (approximately 10^{-2} torr), the yield is 1.8-2.0 g (78-87%).

Anal. Calcd. for $C_{20}H_{28}Cl_4Ru_2$: C, 39.2; H, 4.6; Cl, 23.2; MW 612. Found: C, 39.4; H, 4.5; Cl, 23.3; MW (osmometry in $CHCl_3$) 604.

Properties

The compound is soluble in chloroform and dichloromethane and sparingly soluble in methanol, acetone, and tetrahydrofuran. It is almost insoluble in aromatic and petroleum solvents. Solutions and solid are air stable. The compound melts at 200°. (The checker reports 209-230°, depending on rate of heating.) The 100-MHz ^1H nmr spectrum in $CDCl_3$ (TMS internal standard) shows resonances at δ 1.26 (doublet, $CHCH_3$, J = 7 Hz), 2.13 (singlet, CH_3), 2.88 (septet, $CHCH_3$), and 4.60-4.72 (doublet of doublets, C_6H_4, J = 6 Hz). The far-IR spectrum (Nujol) shows medium-strong bands at 292, 260, and 250(sh) cm^{-1} due to Ru–Cl stretching vibrations.

B. DI-μ-CHLORO-BIS[CHLORO(η^6-HEXAMETHYL-BENZENE)RUTHENIUM(II)]

$$[RuCl_2(\eta^6\text{-}C_{10}H_{14})]_2 + 2C_6(CH_3)_6 \longrightarrow [RuCl_2[\eta^6\text{-}C_6(CH_3)_6]]_2 + 2C_{10}H_{14}$$

Procedure

A mixture of the *p*-cymene complex $[RuCl_2(\eta^6\text{-}C_{10}H_{14})]_2$ (1.0 g, 1.63 mmole) and hexamethylbenzene (10 g, excess) is heated to 180-185° with magnetic stirring for 2 hours. The reaction may be carried out in an open or stoppered flask in an oil bath (an inert atmosphere is not necessary). The crystals of hexamethylbenzene which sublime to the upper walls of the flask are periodically scraped down into the melt. The melt is allowed to come to room temperature, and the solid is broken up and transferred to a pad of Filter Aid. *p*-Cymene and some of the excess of hexamethylbenzene are removed by washing with diethyl ether or hexane; the residual hexamethylbenzene is best removed by sublimation ($40°/10^{-1}$ torr). The solid is washed through the Filter Aid with chloroform or dichloromethane until the washings are colorless (approximately 200 mL is required). The compound is crystallized by addition of hexane and evaporation of the orange-red solution. The yield is 0.87 g (80%).

Anal. Calcd. for $C_{24}H_{36}Cl_4Ru_2$: C, 43.1; H, 5.4; Cl, 21.2; MW 669. Found: C, 43.3; H, 5.5; Cl, 21.9; MW (osmometry) 690.

Properties

The reddish-brown solid is similar in appearance and solubility to the *p*-cymene complex and is air stable as a solid and in solution. It melts at 270°. The 100 MHz ^1H nmr spectrum in $CDCl_3$ (TMS internal standard) shows a singlet due to the methyl protons at δ 2.03. The IR spectrum shows strong bands at 299 and 258 cm^{-1} (broad) due to Ru−Cl stretching vibrations.

C. BIS(η^2-ETHYLENE)(η^6-HEXAMETHYLBENZENE)RUTHENIUM(0)

$$[RuCl_2[\eta^6\text{-}C_6(CH_3)_6]]_2 + 4C_2H_4 + 2C_2H_5OH + 2Na_2CO_3 \longrightarrow$$

$$2Ru[\eta^6C_6(CH_3)_6](\eta^2\text{-}C_2H_4)_2 + 2CH_3CHO + 4NaCl + 2H_2O + 2CO_2$$

All manipulations must be carried out in an inert atmosphere using degassed solvents.[9] To a 25-mL, round-bottom flask fitted with a nitrogen inlet, reflux condenser, and magnetic stirring bar is added the di-chloro(η^6-hexamethylbenzene)ruthenium dimer (0.2 g, 0.3 mmole). Under a counterstream of ethylene, anhydrous sodium carbonate (0.2 g) and ethanol (15 mL) are added. The mixture is stirred and heated under reflux under a slow flow of ethylene for 2 hours; the solution initially turns deep red and finally becomes brown. After cooling to room temperature, solvent is stripped in vacuo and the residue is extracted with four 5 mL portions of hexane. The filtered extract is concentrated under reduced pressure to about 5 mL and cooled to $-78°$. The off-white needles are washed by decantation with two 2-mL portions of cold isopentane (2-methylbutane) and are dried in vacuo. The yield is 0.07 g (37%).

Anal. Calcd. for $C_{16}H_{26}Ru$: C, 60.2; H, 8.2; MW 320. Found: C, 60.2; H, 8.1; MW (mass spectrometry) 320 (^{102}Ru).

Properties

The compound is an air-sensitive solid which should be stored under an inert atmosphere in a refrigerator. It dissolves in all common organic solvents giving very air-sensitive solutions. The 100-MHz ^1H nmr spectrum in C_6D_6 (internal TMS) exhibits two multiplets at δ 1.04 and 1.48 due to the protons of coordinated ethylene and a singlet at δ 1.70 due to the methyl protons of hexamethylbenzene. The IR spectrum (CsI disk) shows a weak band at 1480 cm^{-1} which may be due to a mixed C=C stretching/CH$_2$ deformation mode of coordinated ethylene.

D. (η⁴-1,3-CYCLOHEXADIENE)(η⁶-HEXAMETHYL-BENZENE)RUTHENIUM(0)

$$[RuCl_2[\eta^6\text{-}C_6(CH_3)_6]]_2 + 2C_6H_8 + 2C_2H_5OH + 2Na_2CO_3 \longrightarrow$$

$$2Ru[(\eta^6\text{-}C_6(CH_3)_6](\eta^4\text{-}C_6H_8) + 2CH_3CHO + 4NaCl + 2H_2O + CO_2$$

All manipulations must be carried out in an inert atmosphere using degassed solvents. To a 25-mL, round-bottom flask fitted with a nitrogen inlet, reflux condenser, and magnetic stirring bar is added the di-chloro(η⁶-hexamethylbenzene)ruthenium dimer (0.2 g, 0.3 mmole), anhydrous sodium carbonate (0.2 g), 1,3-cyclohexadiene (Chemsampco, 4692 Kenny Road, Columbus, OH 43320, formerly Chemical Samples Co.; or Ega Chemie, 7924 Steinheim/Albuch, W. Germany) (1 mL, 0.8 g, 1.0 mmole) and ethanol or 2-propanol (15 mL). Slightly lower yields are obtained if 1,4-cyclohexadiene, prepared by Birch reduction of benzene, is used instead of 1,3-cyclohexadiene. The mixture is stirred and heated under reflux for 2.5 hours, during which time the suspended dimer dissolves and the solution turns yellow. Volatile liquids are removed in vacuo, and the residue is extracted with four 5-mL portions of hexane. The filtered extract is concentrated in vacuo and cooled to $-78°$. The pale-yellow crystals are washed by decantation with two 2-mL portions of cold isopentane (2-methylbutane) and dried in vacuo. The yield is 0.13 g (63%).

Anal. Calcd. for $C_{18}H_{26}Ru$: C, 62.8; H, 7.6; MW 344. Found: C, 63.2; H, 7.2; MW (mass spectrometry) 344 (^{102}Ru).

Properties

These are similar to those of the corresponding bis(ethylene) complex. The 100-MHz ^{1}H nmr spectrum in C_6D_6 (internal TMS) shows a pair of multiplets at δ 2.41 and 4.41 due to the outer and inner coordinated diene protons, respectively, a multiplet at δ 1.71 due to the methylene protons, and a singlet at δ 1.95 due to the methyl protons of hexamethylbenzene.

References

1. G. Winkhaus and H. Singer, *J. Organometal. Chem.*, 7, 487 (1967).
2. R. A. Zelonka and M. C. Baird, *Can. J. Chem.*, 50, 3063 (1972).
3. M. A. Bennett and A. K. Smith, *J. Chem. Soc., Dalton Trans.*, 233 (1974).
4. I. Ogata, R. Iwata and Y. Ikeda, *Tetrahedron Lett.*, 3011 (1970). R. Iwata and I. Ogata, *Tetrahedron*, 29, 2753 (1973).
5. A. G. Hinze, *Recl. Trav. Chim. Pays-Bas*, 92, 542 (1973).
6. M. A. Bennett, T-N. Huang, A. K. Smith and T. W. Turney, *J. Chem. Soc. Chem. Comm.*, 582 (1978).
7. M. A. Bennett, T-N. Huang and T. W. Turney, *J. Chem. Soc. Chem. Comm.*, 312 (1979).
8. For example: R. A. Zelonka and M. C. Baird, *J. Organometal. Chem.*, 44, 383 (1972). R. H. Crabtree and A. J. Pearman, *ibid.*, 141, 325 (1977). D. R. Robertson, T. A. Stephenson and T. Arthur, *ibid.*, 162, 121 (1978). M. A. Bennett, T. W. Matheson, G. B. Robertson, W. L. Steffen and T. W. Turney, *J. Chem. Soc. Chem. Comm.*, 32 (1979). M. A. Bennett and T. W. Matheson, *J. Organometal. Chem.*, 153, C25 (1978). *Idem.*, *ibid.*, 175, 87 (1979). H. Werner and R. Werner, *Angew. Chem. Int. Ed. Engl.*, 17, 683 (1978). *Idem.*, *J. Organometal. Chem.*, 174, C63, C67 (1979).
9. D. F. Shriver, *The Manipulation of Air-Sensitive Compounds*, McGraw-Hill, New York, 1969, Chap. 7.

17. SOME η^5-CYCLOPENTADIENYLRUTHENIUM(II) COMPLEXES CONTAINING TRIPHENYLPHOSPHINE

Submitted by M. I. BRUCE,* C. HAMEISTER,* A. G. SWINCER,* and R. C. WALLIS*
Checked by S. D. ITTEL†

The chemistry of $RuCl(PPh_3)_2(\eta^5\text{-}C_5H_5)$ differs markedly from that of the corresponding dicarbonyl complex, $RuCl(CO)_2(\eta^5\text{-}C_5H_5)$.[1] The triphenylphosphine complex was first described by Gilbert and Wilkinson,[2] who obtained it

*Department of Physical and Inorganic Chemistry, University of Adelaide, Adelaide, South Australia 5001.
†Central Research and Development Department, E.I. du Pont de Nemours and Co., Wilmington, DE 19898.

from a 2-day reaction between $RuCl_2(PPh_3)_3$ and cyclopentadiene in benzene. An improved synthesis was described later,[3] from $RuCl_2(PPh_3)_3$ and thallium(I) cyclopentadienide. Both reactions are difficult to employ if large amounts of this complex are required; the competing dimerization of $RuCl_2(PPh_3)_3$ to $[RuCl_2(PPh_3)_2]_2$ reduces yields in the first reaction, especially at temperatures above 20°, while the bulk of the thallium compounds makes the second reaction less convenient. The present method employs a slight modification of the one-pot reaction described earlier.[4]

The most significant differences in the chemistries of $RuCl(PPh_3)_2(\eta^5\text{-}C_5H_5)$ and $RuCl(CO)_2(\eta^5\text{-}C_5H_5)$ are the ready displacement of chloride from the former complex by neutral ligands, L, to form cationic complexes $[RuL(PPh_3)_2\text{-}(\eta^5\text{-}C_5H_5)]^+$ [3,5] and of a triphenylphosphine ligand, as found in the reactions of the hydrido or alkyl complexes.[3] In methanol, for example, the equilibrium

$$RuCl(PPh_3)_2(\eta^5\text{-}C_5H_5) + MeOH \longrightarrow [Ru(MeOH)(PPh_3)_2(\eta^5\text{-}C_5H_5)]^+ + Cl^-$$

lies predominantly to the right, and salts with large anions can be readily isolated. Several novel unsaturated ligands have been made by reactions of alkynes with the hydride or alkyls, including $1,3,4\text{-}\eta^3$-butadienyl,[7] allenyl,[8] cumulenyl,[9] and η^5-pentadienyl[9] species. Terminal acetylenes (1-alkynes) react with the chloride affording complexes containing substituted vinylidene ligands, which are readily deprotonated to the corresponding η^1-acetylides on treatment with bases, such as sodium carbonate, butyllithium, or even alumina.[10] Below are given detailed preparations of $RuCl(PPh_3)_2(\eta^5\text{-}C_5H_5)$, the phenylvinylidene complex $[Ru(C:CHPh)(PPh_3)_2(\eta^5\text{-}C_5H_5)]PF_6$, and the η^1-phenylethynyl derivative, $Ru(C\equiv CPh)(PPh_3)_2(\eta^5\text{-}C_5H_5)$.

A. CHLORO(η⁵-CYCLOPENTADIENYL)BIS(TRIPHENYLPHOS-PHINE)RUTHENIUM(II), RuCl(PPh₃)₂(η⁵-C₅H₅)

$$RuCl_3 + 2PPh_3 + C_5H_6 \longrightarrow RuCl(PPh_3)_2(\eta^5\text{-}C_5H_5) + \cdots$$

Procedure

The reaction is carried out in a 2-L, two-neck, round-bottom flask equipped with a 500-mL dropping funnel and a reflux condenser topped with a nitrogen bypass. The apparatus is purged with nitrogen. Triphenylphosphine (21.0 g, 0.08 mole) is dissolved in 1 L of ethanol by heating. (If the solution is not clear, it should be filtered before proceeding further.) Hydrated ruthenium trichloride (5.0 g, 0.02 mole) is dissolved in ethanol (100 mL) by bringing the mixture to the boil and then allowing the solution to cool. Freshly distilled cyclopentadiene

(10 mL, 8.0 g, 0.12 mole) is added to the ruthenium trichloride solution, and the mixture is transferred to the dropping funnel. The dark-brown solution is then added to the triphenylphosphine solution over a period of 10 minutes while maintaining the temperature at the reflux point. After the ruthenium trichloride/cyclopentadiene solution has been added, the mixture has a dark-brown color, which after 1 hour has lightened to a dark red-orange. The solution, which can now be exposed to air, is filtered quickly while hot and cooled overnight at $-10°$. (The product may also be isolated as fine crystals by evaporating this solution to about one-third the volume on a rotary evaporator. Subsequent work-up is as described below.) Orange crystals separate, leaving a pale yellow-orange supernatant.

The crystals are collected on a sintered-glass filter, washed with ethanol (4 × 25 mL) and with light petroleum (4 × 25 mL), and dried in vacuo. Yield ca. 14 g, 90-95%. The yield depends on the composition of the ruthenium trichloride used. Although nominally a trihydrate, commercially available hydrated ruthenium trichloride contains varying amounts of water. The yield quoted by the submitter was obtained using material of approximate composition $RuCl_3 \cdot 2H_2O$. This reaction has been run successfully using material containing up to 4 molecules of water per $RuCl_3$ unit, in which case ca. 12 g complex was obtained. The checker obtained 12.2 g from $RuCl_3 \cdot 3.2H_2O$.

Anal. Calcd. for $RuCl(PPh_3)_2(\eta^5\text{-}C_5H_5)$: C, 67.7; H, 4.8. Found: C, 67.8; H, 5.3%.

Properties

The complex forms orange crystals mp 130-133° (dec., sealed tube) (checker 134-136°) which are stable in air for prolonged periods. It is insoluble in light petroleum and water, slightly soluble in cold methanol or ethanol, diethyl ether, or cyclohexane, more soluble in chloroform, carbon tetrachloride, dichloromethane, carbon disulfide, and acetone, and very soluble in benzene, acetonitrile, and nitromethane. The ^1H nmr spectrum contains a sharp singlet at τ 5.99 for the C_5H_5 protons and a broad signal at τ 2.84 for the aromatic protons (in $CDCl_3$ solution).

B. (η^5-CYCLOPENTADIENYL)(PHENYLVINYLIDENE)BIS(TRIPHENYLPHOSPHINE)RUTHENIUM(II) HEXAFLUOROPHOSPHATE, [Ru(C:CHPh)(PPh$_3$)$_2$(η^5-C$_5$H$_5$)][PF$_6$]

$$RuCl(PPh_3)_2(\eta^5\text{-}C_5H_5) + HC_2Ph + NH_4[PF_6] \longrightarrow$$

$$[Ru(C\text{:}CHPh)(PPh_3)_2(\eta^5\text{-}C_5H_5)][PF_6] + NH_4Cl$$

The reaction between a terminal acetylene and $RuCl(PPh_3)_2(\eta^5\text{-}C_5H_5)$ in methanol gives a deep-red solution, which on addition of a large anion such as tetraphenylborate, tetrafluoroborate, or hexafluorophosphate affords salts of cationic substituted vinylidene-ruthenium complexes.[10] Examples of complexes containing the parent vinylidene, and disubstituted vinylidene ligands, have also been obtained. Other metals, such as molybdenum,[11] manganese,[12] rhenium,[13] iron,[14] and osmium,[15] are known to form vinylidene complexes, and they are important intermediates in reactions of 1-alkynes with several reagents.

Procedure

A mixture of $RuCl(PPh_3)_2(\eta^5\text{-}C_5H_5)$ (5.0 g, 6.9 mmole) and ammonium hexa-fluorophosphate (1.14 g, 7.0 mmole) is suspended in methanol (500 mL) contained in a 1-L, two-necked flask equipped with a reflux condenser and nitrogen bypass. Phenylacetylene (0.8 g, 7.7 mmole) is then added dropwise to the mixture, which is kept strictly under nitrogen and then heated at reflux tempera-ture for 30 minutes, after which time all the $RuCl(PPh_3)_2(\eta^5\text{-}C_5H_5)$ has dissolved, and the solution is dark red. After cooling, the solution is stable in air for short periods. The solution is filtered to remove any solids and is then evaporated to dryness (rotary evaporator) to give a dark-red solid. The residue is extracted with dichloromethane (2 × 15 mL), and any solid matter (mainly ammonium chloride) that may be present is removed by filtration. The dark-red filtrate is then allowed to drip slowly (2 drops/sec) into vigorously stirred diethyl ether (500 mL). As each drop enters the diethyl ether, a pink flocculent precipitate is formed. The suspension is vigorously stirred for 5 minutes more, and the solid is collected on a sintered-glass funnel, washed with diethyl ether (2 × 25 mL) and petroleum ether (2 × 25 mL) and dried in air. The complex may be purified if necessary by reprecipitation from a dichloromethane solution by addition to diethyl ether in the manner described above. Yield 5.97 g (92.5%).

Anal. Calcd. for $C_{49}H_{41}F_6P_3Ru$: C, 62.7; H, 4.4. Found: C, 62.1; H, 4.4.

Properties

The phenylvinylidene complex forms a reddish-pink powder mp 105-110° (dec., sealed tube). The complex is insoluble in light petroleum and diethyl ether but soluble in benzene, methanol, ethanol, tetrahydrofuran, acetonitrile, dichloro-methane, chloroform, and acetone. The infrared spectrum contains bands at 1620 and 1640 cm^{-1}, one of which is assigned to $\nu(C=C)$ (the other arises from the phenyl groups). The ^1H nmr spectrum has resonances at τ 4.72s (C_5H_5), 4.57t (CHPh) and 2.65 and 2.76m (Ph). The ^{13}C nmr spectra of vinylidene complexes contain a characteristic signal at very low field, which is assigned to the metal-bonded carbon; for the phenylvinylidene complex described above, this resonance occurs at 358.9 ppm. Other signals are found at 95.1 s (C_5H_5), 119.6 s (:CHPh), and 126.5-135.2 m (Ph).

The complex is quite reactive, and the hydrogen on the β-carbon is readily removed by bases to yield the corresponding acetylide. It reacts with methanol to give a benzyl(methoxy)carbene complex, and with oxygen to give a metal carbonyl derivative.[15]

C. (η^5-CYCLOPENTADIENYL)(PHENYLETHYNYL)BIS(TRI-PHENYLPHOSPHINE)RUTHENIUM(II), $Ru(C_2Ph)(PPh_3)_2(\eta^5-C_5H_5)$

$$RuCl(PPh_3)_2(\eta^5\text{-}C_5H_5) + MeOH \rightleftharpoons [Ru(MeOH)(PPh_3)_2(\eta^5\text{-}C_5H_5)]^+ + Cl^-$$

$$[Ru(MeOH)(PPh_3)_2(\eta^5\text{-}C_5H_5)]^+ + HC_2Ph \longrightarrow$$

$$[Ru(C:CHPh)(PPh_3)_2(\eta^5\text{-}C_5H_5)]^+ + MeOH$$

$$[Ru(C:CHPh)(PPh_3)_2(\eta^5\text{-}C_5H_5)]^+ + NaOMe \longrightarrow$$

$$Ru(C_2Ph)(PPh_3)_2(\eta^5\text{-}C_5H_5) + Na^+ + MeOH$$

Although many transition metal complexes containing η^1-bonded substituted acetylides are known, few are available in more than moderate yields via conventional reactions of metal halides with an anionic alkynyl compound of an alkali metal,[16] magnesium,[16] or copper(I)[17] or by dehydrohalogenation in a reaction between the metal halide and a 1-alkyne.[18] More recently, reactions between many 1-alkynes and $RuCl(PPh_3)_2(\eta^5\text{-}C_5H_5)$ have been shown to give cationic vinylidene complexes, which are readily deprotonated to give the corresponding substituted η^1-acetylides.[10] The synthesis of the phenylethynyl derivative is typical; the intermediate phenylvinylidene complex is not isolated.

Procedure

$RuCl(PPh_3)_2(\eta^5\text{-}C_5H_5)$ (4.5 g, 6.2 mmole) is suspended in methanol (400 mL) in a 1-L, two-necked flask equipped with a reflux condenser and a nitrogen bypass. Phenylacetylene (1.0 g, excess) is then added dropwise to the suspension, and the mixture is then heated at reflux temperature for 20-30 minutes. The chloride gradually dissolves to form a cherry-red solution. After cooling, sodium (0.6 g, 0.026 mole) is added in small pieces whereupon a yellow crystalline precipitate separates as the sodium dissolves. The solid is filtered from the colorless to pale yellow-green mother liquor, washed with methanol (4 × 10 mL) and petroleum ether (4 × 10 mL), and dried in vacuo. Yield 4.7 g (96%).

Anal. Calcd. for $C_{49}H_{40}P_2Ru$: C, 74.3; H, 5.1; P, 7.8. Found: C, 73.7; H, 5.1; P, 7.8.

Properties

The phenylethynyl complex forms lemon-yellow crystals mp 205° (dec.) (checker, 200-205°), which are insoluble in light petroleum, diethyl ether, methanol, and ethanol but soluble in dichloromethane, chloroform, and tetrahydrofuran. The infrared spectrum has a sharp $\nu(C\equiv C)$ band at 2068 cm^{-1}, and the ^{1}H nmr spectrum contains resonances at τ 2.52m, 2.90m (35H, Ph) and 5.67s (5H, C_5H_5). The mass spectrum contains a parent ion centered on m/e 792 (calcd. for ^{102}Ru, 792).

As with other complexes containing the $Ru(PPh_3)_2(\eta^5\text{-}C_5H_5)$ moiety, one of the triphenylphosphine ligands may be replaced by other ligands such as CO, $P(OMe)_3$, or Bu^{t}NC. Addition of H$^+$ (using HPF$_6$ or HBF$_4$) or R$^+$ (using $R_3O^+PF_6^-$) affords the corresponding vinylidene complexes.[15] The complex has also been used to make η^2-alkyne complexes containing copper, such as $Ru(C_2PhCuCl)(PPh_3)_2(\eta^5\text{-}C_5H_5)$, originally obtained from $RuCl(PPh_3)_2(\eta^5\text{-}C_5H_5)$ and CuC_2Ph, and the heteronuclear cluster complex $Fe_2Ru(C_2Ph)(CO)_6\text{-}(PPh_3)(\eta^5\text{-}C_5H_5)$.[19,20]

References

1. M. I. Bruce and A. G. Swincer, *Organomet. React. Synth.*, 7, 000 (1982).
2. J. D. Gilbert and G. Wilkinson, *J. Chem. Soc., A*, 1749 (1969).
3. T. Blackmore, M. I. Bruce, and F. G. A. Stone, *J. Chem. Soc., A*, 2376 (1971).
4. M. I. Bruce and N. J. Windsor, *Aust. J. Chem.*, 30, 1601 (1977).
5. G. S. Ashby, M. I. Bruce, I. B. Tomkins, and R. C. Wallis, *Aust. J. Chem.*, 32, 1003 (1979).
6. R. J. Haines and A. L. du Preez, *J. Organometal. Chem.*, 84, 357 (1975).
7. T. Blackmore, M. I. Bruce, and F. G. A. Stone, *J. Chem. Soc. Dalton*, 106 (1974).
8. M. I. Bruce, R. C. F. Gardner, J. A. K. Howard, F. G. A. Stone, M. Welling, and P. Woodward, *J. Chem. Soc., Dalton*, 621 (1977).
9. M. I. Bruce, R. C. F. Gardner, and F. G. A. Stone, *J. Chem. Soc. Dalton*, 906 (1979).
10. M. I. Bruce and R. C. Wallis, *J. Organometal. Chem.*, 161, C1 (1978). *Aust. J. Chem.*, 32, 1471 (1979).
11. R. B. King and M. S. Saran, *J. Am. Chem. Soc.*, 94, 1784 (1972). *Ibid.*, 95, 1811 (1973).
12. A. B. Antonova, N. E. Kolobova, P. V. Petrovsky, B. V. Lokshin, and N. S. Obezyuk, *J. Organometal. Chem.*, 137, 55 (1977).
13. N. E. Kolobova, A. B. Antonova, O. M. Khitrova, M. Y. Antipin, and Y. T. Struchkov, *J. Organometal. Chem.*, 137, 69 (1977).
14. O. S. Mills and A. D. Redhouse, *J. Chem. Soc., A*, 1282 (1968). J. M. Bellerby and M. J. Mays, *J. Organometal. Chem.*, 117, C21 (1976). D. Mansuy, M. Lange, and J. C. Chottard, *J. Am. Chem. Soc.*, 100, 3213 (1978).
15. M. I. Bruce, A. G. Swincer, and R. C. Wallis, *J. Organometal. Chem.*, 171, C5 (1979).
16. M. L. H. Green, *Organometallic Compounds*, 3rd ed., Vol. 2, G. E. Coates, M. L. H. Green and K. Wade (eds.), Methuen, London, 1968, p. 276.
17. M. I. Bruce, R. Clark, J. Howard, and P. Woodward, *J. Organometal. Chem.*, 42, C107 (1972); O. M. Abu Salah and M. I. Bruce, *J. Chem. Soc., Dalton*, 2302 (1974).
18. P. J. Kim, H. Masai, K. Sonogashira, and N. Hagihara, *Inorg. Nuclear Chem. Lett.*, 6,

181 (1970). K. Sonogashira, T. Yatake, Y. Tohda, and N. Hagihara, *J. Chem. Soc. Chem. Comm.*, 291 (1977).
19. O. M. Abu Salah, M. I. Bruce, R. E. Davis, and N. V. Raghavan, *J. Organometal. Chem.*, **64**, C48 (1974).
20. O. M. Abu Salah and M. I. Bruce, *J. Chem. Soc., Dalton*, 2351 (1975).

18. CYCLOPENTADIENYL COMPLEXES OF TITANIUM(III) AND VANADIUM(III)

Submitted by L. E. MANZER*
Checked by E. A. MINTZ† and T. J. MARKS†

The chlorodicyclopentadienyl complexes of Ti(III) and V(III) are useful synthetic reagents for the synthesis of a variety of paramagnetic organometallic and coordination compounds.[1-4] $TiCl(\eta^5\text{-}C_5H_5)_2$ has been prepared a number of ways, including reduction of $TiCl_2(\eta^5\text{-}C_5H_5)_2$ with zinc dust[5] and reaction of $TiCl_3$ with $Mg(C_5H_5)_2$.[6] Previous routes to $VCl(\eta^5\text{-}C_5H_5)_2$ include the reaction of VCl_4 with $Na(C_5H_5)$[7,8] and the oxidation of $V(\eta^5\text{-}C_5H_5)_2$ with HCl,[8,9] $PhCH_2Cl$,[9] or CH_3Cl.[9] These methods suffer from either low yields or the use of nonreadily available reagents. The procedures described below provide high yields of $MCl(C_5H_5)_2$ (M = Ti, V) using readily available reagents. $Tl(C_5H_5)$ should be sublimed prior to use. All solvents were dried over 4-A molecular sieves and purged with nitrogen prior to use.

■ **Caution.** *All reactions and manipulations should be performed under an atmosphere of dry nitrogen, either in a dry box or using Schlenk-tube techniques. Thallium compounds are extremely toxic and should be handled with care.*

A. CHLOROBIS(η^5-CYCLOPENTADIENYL)TITANIUM(III)

$$TiCl_3 + 2\ Tl(C_5H_5) \xrightarrow{\ C_4H_8O\ } TiCl(\eta^5\text{-}C_5H_5)_2 + 2\ TlCl$$

■ **Caution.** *See the note above concerning the use of inert atmosphere. $TiCl_3$ as a dry powder is pyrophoric in air.*

*Central Research and Development Dept., E. I. du Pont de Nemours & Co., Experimental Station, Wilmington, DE 19898. Current address: Petrochemicals Dept., E. I. duPont de Nemours & Co., Victoria, TX 77901.
† Dept. of Chemistry, Northwestern University, Evanston, IL 60201.

Procedure

A single-necked, round-bottomed, 100-mL flask equipped with a reflux condenser and magnetic stirring bar is charged with 2.0 g (13 mmole) titanium chloride, 30 mL dry tetrahydrofuran, and 6.99 g (26 mmole) thallium(I) cyclopentadienide. The solution is heated to reflux for 1 hour, cooled to room temperature, and filtered through a medium frit. The precipitated thallium(I) chloride is washed with tetrahydrofuran until the washings are colorless. The filtrate and washings are combined and stripped by rotary evaporation. Yield 2.59 g of green-yellow crystals (93.8%).

Anal. Calcd. for $C_{10}H_{10}ClTi$: C, 56.25; H, 4.72; Cl, 16.60. Found: C, 56.00; H, 4.86; Cl, 16.05.

Properties

The product is a very air-sensitive, green-yellow crystalline solid. It does not melt below 250°. Its infrared spectrum contains a strong, sharp peak at 1015 cm^{-1} and strong, broad peaks at 795 and 815 cm^{-1}.

B. CHLOROBIS(η^5-CYCLOPENTADIENYL)VANADIUM(III)

$$VCl_3 + 2Tl(C_5H_5) \xrightarrow{C_4H_8O} VCl(\eta^5\text{-}C_5H_5)_2 + 2TlCl$$

■ **Caution.** *See the note above concerning the use of inert atmosphere.*

Procedure

A single-necked, round-bottomed, 100-mL flask equipped with a reflux condenser and a magnetic stirring bar is charged with 2.06 g (13 mmole) vanadium trichloride, 60 mL anhydrous tetrahydrofuran, and 7.10 g (26 mmole) thallium(I) cyclopentadienide. The solution is then heated at reflux for 2 hours, cooled to room temperature, and filtered through a medium frit. The precipitated thallium(I) is washed with tetrahydrofuran until the washings are colorless. The filtrate and washings are combined and stripped by rotary evaporation. The resulting blue-black solid is dissolved in a minimum amount of dichloromethane, and the suspension is filtered through a fine frit. The dichloromethane is then removed by rotary evaporation to give blue-black crystals. Yield 2.76 g (97.5%).

Anal. Calcd. for $C_{10}H_{10}ClV$: C, 55.46; H, 4.65; Cl, 16.37. Found: C, 56.38; H, 4.91, Cl, 16.24.

Properties

The product is an extremely air- and moisture-sensitive blue-black, crystalline solid. It has a melting point of 203-205°. Its infrared spectrum contains two strong, sharp peaks at 1020 and 1010 cm^{-1} and a strong, broad peak at 810 cm^{-1}. It is soluble in THF and CH_2Cl_2 and insoluble in hydrocarbons. It readily sublimes at 10^{-4} torr and 100°.

References

1. P. C. Wailes, R. S. P. Coutts, and H. Weigold, *Organometallic Chemistry of Titanium, Zirconium and Hafnium*, Academic Press, New York, 1974.
2. H. Bowman and J. Teuben, *J. Organometal. Chem.*, **110**, 327 (1976).
3. J. H. Teuben and H. J. De Liefde Meijer, *J. Organometal. Chem.*, **46**, 313 (1972).
4. L. E. Manzer, *J. Am. Chem. Soc.*, **99**, 276 (1977).
5. M. L. H. Green and C. R. Lucas, *J. Chem. Soc., Dalton Trans.*, 1000 (1972).
6. A. F. Reid and P. C. Wailes, *Aust. J. Chem.*, **18**, 9 (1965).
7. S. Vigoureux and P. Keizel, *Chem. Ber.*, **93**, 701 (1960). *Chem. Abstr.*, **54**, 15347 (1960).
8. H. J. De Liefde Meijer, M. H. Jansen, and G. J. M. Van der Kerk, *Chem. Ind.* (London), 119 (1960).
9. H. J. De Liefde Meijer, M. H. Jansen, and G. J. M. Van der Kerk, *Recl. Trav. Chim. Pays-Bas*, **80**, 831 (1961).
10. T. L. Brown and J. A. Ladd, *Adv. Organometal. Chem.*, **2**, 373 (1964).

19. DICHLORO(ETHENE)(N,N,N',N'-TETRAETHYL-1,2-ETHANEDIAMINE)PLATINUM(II) AND RELATED COMPLEXES

Submitted by A. DE RENZI,* A. PANUNZI,* and M. SCALONE*
Checked by JOHN R. WEIR† and RICHARD F. HECK†

A very limited number of examples of platinum(II) complexes, in which ethylene is coordinated with 4 other donor atoms, are known.[1-4] Five-coordinated platinum(II) complexes of the type [PtCl$_2$(C$_2$H$_4$)(diamine)] (diamine = N,N'-substituted-1,2-ethanediamine) were formed[4] by allowing the Zeise's salt to react with the diamine in methanol solution. The N,N,N',N'-tetramethyl-1,2-ethanediamine derivative has been reported[1] to be very unstable in solution as well as in the solid state. The detailed synthesis of [PtCl$_2$(C$_2$H$_4$){(C$_2$H$_5$)$_2$NCH$_2$-CH$_2$N(C$_2$H$_5$)$_2$}] is reported here. Additional five-coordinated ethylenediamine-

*Istituto Chimico, Università di Napoli, Via Mezzocannone 4, 80134 Napoli, Italy.
†Department of Chemistry, University of Delaware, Newark, DE 19711.

platinum(II) complexes[5] can be prepared by the same procedure if N,N,N',N'-tetraethyl-1,2-ethanediamine is replaced with other 1,2-ethanediamines.

The facile synthesis of these complexes should make them a useful starting material for extensive study of the platinum(II) five-coordination state.

A. DICHLORO(ETHENE)(N,N,N',N'-TETRAETHYL-1,2-ETHANEDIAMINE)PLATINUM(II)

$$K[PtCl_3(C_2H_4)] \cdot H_2O + (C_2H_5)_2NCH_2CH_2N(C_2H_5)_2 \longrightarrow$$

$$[PtCl_2(C_2H_4)|\{(C_2H_5)_2NCH_2CH_2N(C_2H_5)_2\}] + KCl + H_2O$$

Procedure

Potassium trichloro(ethylene)platinate(1–) monohydrate (1.55 g, 4.0 mmole) [*Inorg. Synth.*, **14**, 90 (1973)] and 10 mL methanol are placed in a countercurrent of nitrogen in a 25-mL Schlenk-type flask containing a magnetic stirring bar. The mixture is stirred to dissolve the Zeise's salt and cooled with an ice-water bath. Then N,N,N',N'-tetraethyl-1,2-ethanediamine (Fluka A.G., Buchs, Switzerland) (TEED, 0.76 g, 4.4 mmole) is added with a syringe. A yellow precipitate immediately forms and the suspension is stirred 10 minutes. The product is collected on a Buchner funnel by suction filtration, washed with a 1-mL portion of cold methanol, and dried in vacuo. This crude material, containing potassium chloride, weighs 1.98 g. For purification purposes, the crude product is quickly dissolved in 5 mL cold dichloromethane and gravity filtered. (Recrystallization should be performed as rapidly as possible in order to avoid decomposition in solution.) Upon addition of methanol (5 mL) and evaporative concentration of the orange solution to 5 mL, fine yellow-orange needles form. (The checkers found it necessary to concentrate the solution by evaporation to ~2.5 mL. A yield of ~70% was obtained.) The crystals (1.28 g) are collected on a Buchner funnel, washed with a 1-mL portion of cold methanol, and dried in vacuo. Additional product (0.21 g) may be obtained by further evaporation of the mother liquor. The total yield amounts to 1.49 g (80%).

Anal. Calcd. for $C_{12}H_{28}Cl_2N_2Pt$: C, 30.9; H, 6.0; Cl, 15.2; MW 466.4. Found: C, 31.2; H, 6.2; Cl, 15.1; MW 475 (osmometric, benzene).

B. DICHLORO(ETHENE)(DIAMINE)PLATINUM(II)

$$K[PtCl_3(C_2H_4)] \cdot H_2O + diamine \longrightarrow [PtCl_2(C_2H_4)(diamine)] +$$

$$KCl + H_2O$$

[diamine = N,N'-bis(1-methylethyl)-1,2-ethanediamine*[†] (BMED); N,N'-dimethyl-N,N'-bis(1-methylethyl)-1,2-ethanediamine*[†] (DMMED); (S,S)-N,N'-bis(1-phenylethyl)-1,2-ethanediamine*[†] (BPED); or (R,R)-N,N'-dimethyl-N,N'-bis(1-phenylethyl)-1,2-ethanediamine*[†] (DMPED)]

Procedure

The method of preparation of the five-coordinate platinum(II) complexes of the above diamines is identical to the one used in Section 19-A. All the compounds can be purified by recrystallization from a dichloromethane/methanol mixture. The yields range 65-85% (see Table I).

TABLE I   Yields, Melting Points, and ^{1}H nmr Data for $[PtCl_2(C_2H_4)(RR'NCH_2CH_2NRR')]$ Compounds

R, R'	mp (°C) (dec.)	Yields (%)	$\delta(C_2H_4)$ (ppm)	$^2J(^{195}PtH)$ (Hz)
R,R' = C_2H_5	92-94	80	3.62[a]	72
R = H; R' = CH(CH$_3$)$_2$	120-125	65	3.40[a]	71
R = CH$_3$; R' = CH(CH$_3$)$_2$	100-105	70	3.60[a]	72
R = H; R' = (S)-CH(CH$_3$)C$_6$H$_5$	168-172	83	3.50[b]	—
R = CH$_3$; R' = (R)-CH(CH$_3$)C$_6$H$_5$	133-135	84	3.76[a]	72

[a] Singlet with ^{195}Pt satellite peaks.
[b] AA'BB' multiplet. $^2J_{(195PtH)}$ not evaluable.

Anal. Calcd. for $C_{10}H_{24}Cl_2N_2Pt$ (BMED complex): C, 27.4; H, 5.5; Cl, 16.2; MW 438.3. Found: C, 27.2; H, 5.3; Cl, 16.3; MW 427 (osmometric, benzene). Calcd. for $C_{12}H_{28}Cl_2N_2Pt$ (DMMED complex): C, 30.9; H, 6.05; Cl, 15.2; MW 466.4. Found: C, 30.9; H, 6.0; Cl, 15.1; MW 445 (osmometric, benzene). Calcd. for $C_{20}H_{28}Cl_2N_2Pt$ (BPED complex): C, 42.7; H, 5.0; Cl, 12.6; MW 562.4. Found: C, 42.7; H, 4.9; Cl, 12.7; MW 580 (osmometric, chloroform). Calcd. for

[†]Symmetrically N,N'-substituted 1,2-ethanediamines can be prepared[4] by properly adapting the method of S. Caspe, *J. Am. Chem. Soc.*, **54**, 4457 (1932), by the following general route:

$$2RR'NH + BrCH_2CH_2Br \longrightarrow RR'\overset{+}{N}HCH_2CH_2\overset{+}{N}HRR' + 2Br^- \xrightarrow{KOH}$$

$$RR'NCH_2CH_2NRR' + 2KBr + 2H_2O$$

$C_{22}H_{32}Cl_2N_2Pt$ (DMPED complex): C, 44.75; H, 5.45; Cl, 12.0; MW 590.5. Found: C, 44.7; H, 5.4; Cl, 12.0; MW 580 (osmometric, chloroform).

Properties

All the compounds crystallize in the form of yellow to yellow-orange, air-stable needles. Melting points (with decomposition) are listed in Table I.

Complexes containing TEED, BMED, and DMMED are soluble in dichloromethane, chloroform, acetone, and benzene. They also are moderately soluble in methanol and diethyl ether and insoluble in alkanes. In solutions made from chlorinated solvents or acetone, these complexes decompose within 1-6 hours to an unidentified intermediate. Yellow-green, well-formed crystals of the corresponding [$PtCl_2$(diamine)] complexes separate after 2-3 days at room temperature in a nearly quantitative yield. A remarkably slower decomposition process is observed in benzene solution.

Complexes containing BPED and DMPED are soluble in dichloromethane and chloroform, slightly soluble in benzene and acetone, and insoluble in diethyl ether, methanol, and alkanes. A slow decomposition takes place in solution which can however be neglected, provided the solutions are not aged for longer than 24 hours.

Conductance measurements on fresh solutions of all these compounds clearly show that they are not electrolytes. Infrared spectra have a single, intense absorption band in the 330-340 cm^{-1} region (Pt–Cl stretching). Relevant data from ^1H nmr spectra, recorded in benzene-d_6 solution at 60 MHz and referenced to TMS, are reported in Table I.

Acknowledgments

This work was supported by grants from the National Research Council (CNR).

References

1. L. Maresca, G. Natile, and L. Cattalini, *Inorg. Chim. Acta*, **14**, 79 (1975).
2. L. E. Manzer, *Inorg. Chem.*, **15**, 2354 (1976).
3. L. Maresca, G. Natile, M. Calligaris, P. Delise, and L. Randaccio, *J. Chem. Soc., Dalton Trans.*, **1976**, 2386.
4. A. De Renzi, A. Panunzi, A. Saporito, and A. Vitagliano, *Gazz. Chim. Ital.*, **107**, 549 (1977).
5. A. DeRenzi, B. DiBlasio, A. Saporito, M. Scalone, and A. Vitagliano, *Inorg. Chem.*, **19**, 960 (1980).

20.   1,2-ETHANEDIYLBIS(DIPHENYLPHOSPHINE) COMPLEXES OF IRON

Submitted by S. D. ITTEL* and M. A. CUSHING, JR.*
Checked by R. EISENBERG†

Zerovalent complexes of iron-containing phosphorus ligands are of interest for several reasons. The five-coordinate structures are stereochemically nonrigid on the nmr time scale, resulting in interesting spectroscopic behavior.[1] In addition, they are much more electron rich than their more common carbonyl analogs, giving rise to a variety of interesting oxidative addition reactions, including in some cases the cleavage of carbon-hydrogen bonds.[2,3]

The 1,2-ethanediylbis(diphenylphosphine) (dppe) complexes of iron reported here are readily prepared from relatively inexpensive materials and display the above properties.[3,4] The starting material $Fe(dppe)_2(C_2H_4)$ is prepared by alkylaluminum reduction of $Fe(acac)_3$ (acac = 2,4-pentanedionato) in the presence of dppe. The reaction has been shown to proceed through the intermediate $Fe(dppe)(acac)_2$, which is more conveniently isolated from the reaction of $Fe(dppe)_2C_2H_4$ with 2,4-pentanedione.[5] $Fe(dppe)_2(C_2H_4)$ reacts with acetylenes to form *trans*-hydrido acetylide complexes and with cyclopentadiene to give a cyclopentadienyl iron dppe hydride. $Fe(dppe)_2(C_2H_4)$ also reacts with CO, *t*-BuNC, and phosphorus ligands[6] to give $Fe(dppe)_2L$ species, but these are more conveniently isolated from reactions involving the *ortho*-metallated species **1**:

(1)

*Central Research and Development Department, E. I. du Pont de Nemours and Co., Experimental Station, Wilmington, DE 19898.
†Department of Chemistry, University of Rochester, Rochester, NY 14627.

A. BIS[1,2-ETHANEDIYLBIS(DIPHENYLPHOSPHINE)](ETHENE)IRON

$$Fe(C_5H_7O_2)_3 + 2(C_6H_5)_2PCH_2CH_2P(C_6H_5)_2 + 3Al(OC_2H_5)(C_2H_5) \longrightarrow$$

$$Fe(CH_2CH_2)[(C_6H_5)_2PCH_2CH_2P(C_6H_5)]_2 + 3\ Al(OC_2H_5)(C_5H_7O_2)C_2H_5 +$$

$$C_2H_4 + C_2H_6$$

Procedure

■ **Caution.** *Alkylaluminum compounds, products, and the by-products of this reaction are extremely air sensitive and must be handled in a rigorously oxygen-free atmosphere.*

The entire procedure is performed in an anhydrous, oxygen-free atmosphere, using anhydrous, deoxygenated solvents. Standard techniques for bench-top inert-atmosphere reactions are used throughout.[7] Tris(2,4-pentanedionato)-iron(III) (42.4 g, 120 mmole) and 1,2-ethanediylbis(diphenylphosphine) (dppe) (95.2 g, 240 mmole) are placed in a three-necked, 1-L flask equipped with a mechanical stirrer and nitrogen bubbler. The flask is slowly flushed with nitrogen. A diethyl ether (100 mL) solution of ethoxydiethylaluminum* (65 g, 76 mL, 500 mmole) is cautiously prepared. (This is a highly exothermic reaction which can cause the diethyl ether to boil vigorously.) The alkylaluminum solution is transferred to a pressure-equalizing dropping funnel equipped with a stopcock in the sidearm.

Diethyl ether (400 mL) is syringed into the degassed flask, and the dropping funnel of alkylaluminum is put in place. The flask and its contents are cooled to $-15°$ before the aluminum alkyl solution is added over 45 minutes. (Dry ice is added to a toluene bath as necessary to obtain the desired temperature. A wet ice/methanol bath may also be used but is somewhat more hazardous because of the violent reaction of triethylaluminum with water and methanol.) During the addition, the color changes from orange-red to golden brown Fe(dppe)(acac)$_2$ and finally to brick red. After the addition is complete, the mixture is allowed to warm slowly to room temperature. Stirring is continued for at least 3 hours. The product is collected by vacuum filtration and washed with diethyl ether. It is then resuspended in diethyl ether, warmed and stirred for several minutes, and collected again by vacuum filtration. The material is of analytical purity. The filtrate contains highly reactive ethylaluminum compounds which react violently with water or alcohols and is most conveniently disposed of by incineration.

---

*Triethylaluminum can be substituted for the ethoxydiethylaluminum, but the product is invariably contaminated with Fe(dppe)$_2$H$_2$. The dihydride is only difficultly removed by crystallization and will be carried through subsequent steps in these syntheses.

Alternatively, the mixture may be decomposed by the careful, dropwise addition of 200 mL ethanol to the cooled, stirred solution (much gas is evolved), followed by the cautious addition of water. The yield of unrecrystallized product is 85 g (80%).

Anal. Calcd. for $FeP_4C_{54}H_{52}$: C, 73.6; H, 5.95; P, 14.1; Fe, 6.3. Found: C, 73.7; H, 5.81; P, 15.0; Fe, 7.6; mp 165-166°.

Properties

Bis[1,2-ethanediylbis(diphenylphosphine)] (ethylene)iron is a brick-red powder which is mildly air sensitive and should be stored and used in an inert atmosphere. The complex is only slightly soluble in diethyl ether and more soluble in aromatic solvents and tetrahydrofuran. The $^{31}P\{^1H\}$ nmr spectrum consists of an A_2B_2 spin system: $\delta_A = 94.5$; $\delta_B = 78.1$ ppm, $J_{AB} = 40$ Hz.

The analogous diphosphine complex can be prepared with methylenebis(diphenylphosphine) but not with 1,3-propanediylbis(diphenylphosphine).

B.   {2-[2-DIPHENYLPHOSPHINO)ETHYL] PHENYLPHOSPHINO}PHENYL-*C,P,P'*[1,2-ETHANEDIYLBIS(DIPHENYLPHOSPHINE)]HYDRIDOIRON

$$Fe(C_2H_4)[(C_6H_5)_2PCH_2CH_2P(C_6H_5)_2] \xrightarrow{\Delta} C_2H_4 +$$

Procedure

Bis[(1,2-ethanediylbis(diphenylphosphine)] (ethene)iron (17.6 g, 20 mmole) is suspended in toluene (120 mL) in a three-necked, 250-mL flask equipped with a nitrogen bubbler, reflux condenser, and magnetic stirring bar. The mixture is brought to reflux and maintained at this temperature for 20 minutes. The solution lightens slightly over this time as ethylene is evolved. The hot mixture is filtered and then cooled to room temperature. The volume of solvent is reduced under vacuum to about 25 mL. Pentane (80 mL) is added slowly, and the resultant orange solids are collected by vacuum filtration and washed with

50-mL portions of pentane. They are then dried under vacuum. The yield is 14 g (80%).

Anal. Calcd. for $FeP_4C_{52}H_{48}$: C, 73.2; H, 5.67; P, 14.5; Fe, 6.6. Found: C, 72.7; H, 5.70; P, 15.1; Fe, 7.7; mp 179-180°.

Properties

The *ortho*-metallated product is an orange powder which is mildly air sensitive and should be stored and used in an inert atmosphere. It is generally more soluble than its precursor with increasing solubility in diethyl ether, tetrahydrofuran, and aromatic solvents.

C. BIS[1,2-ETHANEDIYLBIS(DIPHENYLPHOSPHINE)](TRIMETHYL PHOSPHITE)IRON

$$Fe[(C_6H_5)_2PCH_2CH_2P(C_6H_5)_2]H[(C_6H_4)(C_6H_5)PCH_2CH_2P(C_6H_5)_2] +$$

$$P(OCH_3)_3 \longrightarrow Fe[P(OCH_3)_3][(C_6H_5)_2PCH_2CH_2P(C_6H_5)_2]_2$$

Procedure

■ **Caution.** *Trimethyl phosphite has a strongly disagreeable odor. Work in a well-ventilated area.*

A small Schlenk flask equipped with a serum stopper and magnetic stirring bar is charged with the *ortho*-metallated product of the preceding section (2.56 g, 3.0 mmole) and toluene (40 mL). Excess trimethyl phosphite (0.74 g, 0.71 mL, 6.0 mmole) is syringed into the solution which is then stirred at room temperature for 30 minutes. Half of the solvent is removed under vacuum, and then diethyl ether (40 mL) is slowly syringed into the flask causing precipitation of orange-brown solids. The flask is cooled to 0° for 20 minutes, and the product is collected by vacuum filtration. The product is recrystallized by dissolving it in toluene (15 mL) and precipitating it with diethyl ether (40 mL) at 0°. The yield of recrystallized material is 1.8 g (60%).

Anal. Calcd. for $FeP_5O_3C_{55}H_{57}$: C, 67.6; H, 5.89; P, 15.9; Fe, 4.9. Found: C, 67.4; H, 5.91; P, 15.9; Fe, 5.1; mp 146.7°.

Properties

Bis[1,2-ethanediylbis(diphenylphosphine)](trimethyl phosphite)iron is a red-brown, crystalline material which is mildly air sensitive and should be stored and handled in an inert atmosphere. It is five-coordinate and stereochemically nonrigid on the ^{31}P nmr time scale; at room temperature, it is an AB$_4$ spin system, δ_A = 155.2 ppm, δ_B = 86.9 ppm (from phosphoric acid), J_{AB} = 10 Hz; and at $-90°$, it displays a complex AB$_2$CD spin system.

D. [1,2-ETHANEDIYLBIS(DIPHENYLPHOSPHINE)] BIS(2,4-PENTANE-DIONATO)IRON

$$2Fe[(C_6H_5)_2PCH_2CH_2P(C_6H_5)_2]_2(C_2H_4) + 2CH_3COCH_2COCH_3 \longrightarrow$$

$$Fe[(C_6H_5)_2PCH_2CH_2P(C_6H_5)_2](CH_3COCHCOCH_3)_2 +$$

$$Fe[(C_6H_5)_2PCH_2CH_2P(C_6H_5)_2]_2H_2 +$$

$$(C_6H_5)_2PCH_2CH_2P(C_6H_5)_2 + 2C_2H_4$$

$$Fe[(C_6H_5)_2PCH_2CH_2P(C_6H_5)]_2H_2 + 2CH_3COCH_2COCH_3 \longrightarrow$$

$$Fe[(C_6H_5)_2PCH_2CH_2P(C_6H_5)](CH_3COCHCOCH_3)_2 +$$

$$(C_6H_5)_2PCH_2CH_2P(C_6H_5) + 2H_2$$

Procedure

A three-necked, 250-mL flask equipped with nitrogen bubbler, reflux condenser, and magnetic stirring bar is charged under a blanket of inert gas with bis[1,2-ethanediylbis(diphenylphosphine)] (ethylene)iron (5.28 g, 6.0 mmole), toluene (50 mL), and 2,4-pentanedione (6.0 g, 60 mmole). The solution is brought to reflux and maintained for 30 minutes. The mixture lightens from dark red to dark yellow. Cooling the mixture to room temperature and slowly adding diethyl ether (150 mL) result in the formation of long, fibrous needles which are collected by vacuum filtration. The product is washed with diethyl ether and dried under vacuum. The yield is 1.66 g (85%). The material is easily recrystallized from tetrahydrofuran or toluene/diethyl ether.

Anal. Calcd. for FeP$_2$O$_4$C$_{36}$H$_{38}$: C, 66.3; H, 5.87; P, 9.5; O, 9.8. Found: C, 65.9; H, 5.83; P, 9.7; O, 9.6; mp 193-194°.

Properties

The golden, highly crystalline material is air stable. It is paramagnetic having μ_{EFF} = 5.10 B.M. as measured by the nmr technique. It is soluble in tetrahydrofuran, aromatic solvents, and chlorinated solvents.

Analogous complexes can be prepared with other β-diketones. Dibenzoylmethane gives a dark-purple complex using the above procedure. Hexafluoroacetylacetone gives a dark-red crystalline product which is separated from the reaction by-products by virtue of its solubility in pentane.

References

1. A. D. English, S. D. Ittel, C. A. Tolman, P. Meakin, and J. P. Jesson, *J. Am. Chem. Soc.*, **99**, 117 (1977).
2. C. A. Tolman, S. D. Ittel, A. D. English, and J. P. Jesson, *J. Am. Chem. Soc.*, **101**, 1742 (1979).
3. S. D. Ittel, C. A. Tolman, P. J. Krusic, A. D. English, and J. P. Jesson, *Inorg. Chem.*, **17**, 3432 (1978).
4. T. Ikariya and A. Yamamoto, *J. Organometal. Chem.*, **118**, 65 (1976).
5. S. D. Ittel, *Inorg. Chem.*, **16**, 1245 (1977).
6. C. A. Tolman and J. P. Jesson, U.S. Patent 3,997,579. 1976
7. D. F. Shriver, *The Manipulation of Air-Sensitive Compounds*, McGraw-Hill Book Company, New York, 1969.

Chapter Four

COORDINATION COMPOUNDS

21. *trans*-CARBONYLCHLOROBIS(DIMETHYL-PHENYLPHOSPHINE)IRIDIUM(I)

Submitted by L. R. SMITH,* S. M. LIN,* M. G. CHEN,* J. U. MONDAL,* and D. M. BLAKE*
Checked by S. D. LEHR†

The dimethylphenylphosphine complex *trans*-$\{IrCl(CO)[P(CH_3)_2C_6H_5]_2\}$ is very useful in the study of the oxidative addition reaction. In addition to its high reactivity, the ^1H nmr spectra of adducts of the complex are a convenient tool for obtaining stereochemical information about the addition of small molecules to square planar complexes.[1]

$$trans\text{-}\{IrClCO[P(C_6H_5)_3]_2\} + 2\ P(CH_3)_2(C_6H_5) \xrightarrow{(C_2H_5)_2O}$$
$$trans\text{-}\{IrCl(CO)[P(CH_3)_2C_6H_5]_2\} + 2\ P(C_6H_5)_3$$

Procedure

■ **Caution.** *Dimethylphenylphosphine is a foul-smelling, corrosive compound. Toxicity or other health hazards have not been determined. All manipulations*

*Department of Chemistry, Box 19065, The University of Texas at Arlington, Arlington, TX 76019.
†Dept. of Chemistry, Case Western Reserve University, Cleveland, OH 44106.

97

should be carried out in a well-ventilated hood. The following procedure should be carried out using standard techniques for the manipulation of air-sensitive compounds.

The reaction apparatus is a Schlenk system consisting of two 100-mL, two-necked flasks each containing a stirring bar. The flasks are joined via a medium porosity filter tube affixed in one neck of each flask. The second neck is closed with a rubber serum stopper. A dinitrogen flow is passed into the reaction flask and out of the receiving flask via syringe needles in the serum caps. *trans*-$\{IrCl(CO)[P(C_6H_5)_3]_2\}$ is recrystallized[7] from benzene/methanol. Dimethyl-phenylphosphine (0.76 mL, 5.5 mmole) is added via syringe to a suspension of *trans*-$\{IrCl(CO)[P(C_6H_5)_3]_2\}$[2,3] (2.0 g, 2.6 mmole) in dioxygen-free diethyl ether (60 mL). The yellow, suspended solid gradually dissolves as the mixture is stirred for about 2.5 hours. (The checker found that some residue remained even after 24 hours. This was removed by filtration.) The resulting orange solution is filtered through the filter tube into the second flask by inverting the apparatus. The volume of the filtrate is reduced to about 30 mL by evaporation using a flow of dinitrogen (The checker then removed all the solvent by evaporation, washed the residue with methanol, added 30 mL diethyl ether, and continued as indicated.). Dioxygen-free hexane (30 mL) is added to the orange solution and the volume again is reduced to about 30 mL using a dinitrogen flow. The mixture is cooled in a freezer at about $-10°$ for 4 hours. The yellow, crystalline product is collected by filtration, opened to the air, and quickly washed with cold hexane (30 mL) in three portions.

The solid is recrystallized, in the Schlenk apparatus, by dissolving it in 80 mL of 9:1 diethyl ether/dichloromethane, filtering and evaporating the solution to about 40 mL. Hexane (40 mL) is added and the volume again reduced by a dinitrogen flow to 40 mL. Cooling in a freezer for 3 hours causes precipitation of the yellow solid, which is recovered by filtration and washed with 30 mL cold hexane. The recovered yield is 1.0 g (73%), mp 109-112°; IR (Nujol) 1957, $\nu(C\equiv O)$ and 310, $\nu(Ir-Cl)$ cm^{-1}; nmr (CDCl$_3$) 1.93δ t(PCH$_3$), 7.2δ m(PC$_6$H$_5$). (Lit.[4] mp 111-113°, 1959 cm^{-1}, 311 cm^{-1}.)

Properties

The product is a yellow, crystalline solid which reacts with dioxygen rapidly when in solution and slowly when in the solid state. It can be stored in sealed ampuls or tightly closed bottles, under dinitrogen, for an indefinite period. Syntheses of other complexes in the series $P(CH_3)_n(C_6H_5)_{3-n}$ are available in the literature ($n = 3$[5], 1[6], or 0[2]).

References and Notes

1. (a) J. M. Jenkins and B. L. Shaw, *Proc. Chem. Soc.* (London), 291 (1963). (b) *J. Chem. Soc., A*, 770 (1966). (c) J. M. Jenkins, M. S. Lupin, and B. L. Shaw, *J. Chem. Soc., A*,

1787 (1966). (d) A. J. Deeming and B. L. Shaw, *Chem. Comm.*, 751 (1968).

2. J. P. Collman, C. T. Sears, and M. Kubota, *Inorg. Synth.*, **11**, 102 (1968).

3. In our experience, unrecrystallized material gives a very low or no yield of product. Material recrystallized from chloroform also may give erratic results. The checker finds that recrystallization is essential. Addition of a small amount (less than 10% by weight) of $P(C_6H_5)_3$ was found to aid the process.

4. A. J. Deeming and B. L. Shaw, *J. Chem. Soc., A*, 1887 (1968).

5. J. A. Labinger and J. A. Osborn, *Inorg. Synth.*, **18**, 64 (1978).

6. J. P. Collman and J. W. Kang, *J. Am. Chem. Soc.*, **89**, 844 (1967).

22. CARBON DIOXIDE COMPLEXES OF RHODIUM AND IRIDIUM

Submitted by T. HERSKOVITZ*
Checked by C. KAMPE,† H. D. KAESZ,† and WM. SEIDEL*

In recent years, interest has intensified in the chemistry of carbon dioxide, stimulated by the current concern about alternate petrochemical feedstocks. One area under active exploration involves CO_2 activation via coordination to a transition metal complex.[1] Several adducts of CO_2 have been claimed, and two monometallic complexes, with x-ray structures which have been published, are shown below schematically (**1** and **2**).[2,3] We report here two examples of the preparation of 1:1 CO_2 adducts of a series of rhodium and iridium complexes and, relatedly, methods for preparing the 2:1 CO_2:Ir complex **2**.

(1)   (2)

*Central Research & Development Department, E. I. du Pont de Nemours & Co., Experimental Station, Wilmington, Delaware 19898, contribution No. 2942.
†Dept. of Chemistry, University of California, Los Angeles, CA 90024.

A. (CARBON DIOXIDE)BIS[1,2-ETHANEDIYLBIS(DIMETHYLPHOS-
PHINE)]IRIDIUM CHLORIDE[4]

$$Ir_2Cl_2(C_8H_{14})_4 + 4(CH_3)_2PCH_2CH_2P(CH_3)_2* \longrightarrow 2[Ir(dmpe)_2]Cl$$
$$\text{dmpe}$$

$$[Ir(dmpe)_2]Cl + CO_2 \longrightarrow [Ir(dmpe)_2(CO_2)]Cl$$

Procedure

■ **Caution.** *The phosphines and arsines used are spontaneously flammable in air and are toxic.*

The following manipulations are all done in an atmosphere of dry nitrogen using dry, deoxygenated, reagent-grade solvents. Coleman Instrument Grade (99.99% minimum) carbon dioxide is used. Dmpe is commercially available.

To a stirred solution of 3.25 g (3.63 mmole) $Ir_2Cl_2(C_8H_{14})_4$[5] in 100 mL toluene, 2.18 g (14.5 mmole) $(CH_3)_2PC_2H_4P(CH_3)_2$ (dmpe) in 10 mL toluene is added. The mixture is stirred for 30 minutes then filtered, and the resultant solid is washed with toluene and then with pentane. Drying to <0.1 micron affords 2.9 g of a light-orange solid (75% yield), mp (vac.) 252-253° (dec.). (The checkers (C.K. and H. D. K.) report the formation of the cyclometalation product[7] when the procedure is followed without exclusion of light, ν(Ir—H) = 2180 cm^{-1}.)

Anal. Calcd. for $C_{12}H_{32}P_4ClIr$ ([Ir(dmpe)$_2$]Cl): C, 27.30; H, 6.10; P, 23.46; Cl, 6.72; Ir, 36.41. Found: C, 26.93; H, 5.99; P, 23.23; Cl, 6.48; Ir, 36.11.

A stirred suspension of 1.0 g [Ir(dmpe)$_2$]Cl in 20 mL toluene is placed under 15 psig CO_2, resulting within minutes in a suspension of a light-tan solid. After $\geqslant$1 hour, the solid is collected, washed with toluene then hexane, and then dried to <0.1 micron. The resultant air-sensitive, off-white solid, mp (vac.) 196-198° (dec.), exhibits an IR spectrum (mull) with bands at 2180(w), 1640(w), 1550(s), 1230(s), and 765(s) cm^{-1}.

Anal. Calcd. for $C_{13}H_{32}O_2P_4ClIr$(Ir(dmpe)$_2$(CO$_2$)Cl): C, 27.30; H, 5.64; O, 5.60; P, 21.66; Cl, 6.20; Ir, 33.60. Found: C, 27.65; H, 5.76; O, 5.42; P, 21.56; Cl, 6.44; Ir, 33.29.

The same complex can be prepared by bubbling CO_2 for several minutes through a suspension of [Ir(dmpe)$_2$]Cl in diethyl ether or any of the usual hydrocarbon solvents. Alternatively, the orange solid, [Ir(dmpe)$_2$]Cl, within 1 hour affords, quantitatively, white Ir(dmpe)$_2$(CO$_2$)Cl upon exposure to 15 psig carbon dioxide.

---

*The alternative name 1,2-bis(*dim*ethyl*p*hosphino)ethane, dmpe, is commonly used for this ligand.

Properties

The $Ir(dmpe)_2(CO_2)Cl$ complex may be handled in air for minutes as a solid. It may be stored indefinitely under nitrogen and is stable to vacuum. Ethylene does not react with it, but CO and SO_2 displace the bound CO_2. Hydrogen and acetonitrile displace the CO_2 and oxidatively add, affording $[Ir(H)_2(dmpe)_2]Cl$ and $[Ir(H)(CH_2CN)(dmpe)_2]Cl$, respectively. Most solvents that dissolve $Ir(dmpe)_2(CO_2)Cl$ react with it. As a consequence, growing crystals for a structure determination has been unsuccessful. Isotope labeling ($^{13}CO_2$ and $C^{18}O_2$) has shown the bands at 1550(s) and 1230(s) cm^{-1} (mull) to involve the bound CO_2.[4] The weak bands at 2180 and 1640 cm^{-1} are due to a minor impurity.[7]

B. (CARBON DIOXIDE)BIS[o-PHENYLENEBIS(DIMETHYLARSINE)]- RHODIUM CHLORIDE

$$Rh_2Cl_2(C_2H_4)_4 + 4 \quad \underset{\text{(diars)}}{\left[\begin{array}{c} As(CH_3)_2 \\ \\ As(CH_3)_2 \end{array} \right.} \quad \longrightarrow \quad 2[Rh(diars)_2]Cl + 4C_2H_4$$

$$[Rh(diars)_2]Cl + CO_2 \longrightarrow [Rh(diars)_2(CO_2)]Cl$$

Procedure

To a stirred, red solution of 1.00 g (2.6 mmole) $Rh_2Cl_2(C_2H_4)_4$[6] in 25 mL tetrahydrofuran (THF), 2.95 g (10.3 mmole) o-phenylenebis(dimethylarsine) (diars) in 5 mL THF is added dropwise. Filtration after ½ hour affords a light-brown solid which is recrystallized from acetonitrile with diethyl ether. The resultant golden crystals are diethyl ether washed and then vacuum dried, yielding 2.32 g (63% yield). The IR spectrum shows ~2240(vw) cm^{-1} ascribable to $\nu_{C\equiv N}$ of lattice acetonitrile.

Anal. Calcd. for $C_{21}H_{33\frac{1}{2}}N_{\frac{1}{2}}ClAs_4Rh$ ($[Rh(diars)_2]Cl \cdot \frac{1}{2}CH_3CN$): C, 34.50; H, 4.62. Found: C, 34.70; H, 4.63.

A mixture of 0.30 g $[Rh(diars)_2]Cl$ in acetonitrile is placed in a pressure bottle and then evacuated and pressured to 20 psig with CO_2. Overnight the golden solution and solid both become yellow. (To show reversibility: the mixture can be evacuated at 70° for minutes, affording a medium-orange solution. Repressuring with CO_2 causes reversion to a yellow solution.) Filtration after 18 hours yields 0.2 g light-tan solid with IR bands in a Nujol mull at 1620(s), 1205(s), and 1190(sh) cm^{-1}.

Anal. Calcd. for $C_{21}H_{32}O_2ClAs_4Rh$ $[Rh(diars)_2(CO_2)Cl]$: C, 33.43; H, 4.28; O, 4.24. Found: C, 33.74; H, 4.34; O, 5.06.

Properties

The same 1:1 CO_2 adduct is formed upon addition of 20 psig CO_2 to a solution of $[Rh(diars)_2]Cl$ in acetone. This solvent affords a noncrystalline solid, but small crystals are obtained from acetonitrile. A recent x-ray structure determination[8] has shown that this adduct contains η^1-bonded CO_2. $Rh(diars)_2(CO_2)Cl$ has stability properties similar to those of $Ir(dmpe)_2(CO_2)Cl$.

C. CHLORO[(FORMYL-κC-OXY)FORMATO-$\kappa O(2-)$]TRIS(TRIMETHYL-PHOSPHINE)IRIDIUM

$$Ir_2Cl_2(C_8H_{14})_4 + 6P(CH_3)_3 \longrightarrow 2IrCl(C_8H_{14})[P(CH_3)_3]_3 + 2C_8H_{14}$$

$$IrCl(C_8H_{14})[P(CH_3)_3]_3 + xsCO_2 \longrightarrow Ir(C_2O_4)(Cl)[P(CH_3)_3]_3 + C_8H_{14}$$

or

$$Ir_2Cl_2(C_8H_{14})_4 + 8PMe_3 \longrightarrow 2[Ir(PMe_3)_4]Cl + 4C_8H_{14}$$

$$Ir(PMe_3)_4]Cl + xsCO_2 \longrightarrow Ir(C_2O_4)Cl(PMe_3)_3 + PMe_3$$

Procedure

Method 1. To a solution of 6.0 g $Ir_2Cl_2(C_8H_{14})_4{}^5$ (6.7 mmole) in 600 mL toluene, a cold ($-35°$) solution of 3.10 g $P(CH_3)_3$ (40.8 mmole) in 10 mL toluene is added dropwise. After standing at room temperature for 3 days, the solvent is stripped from the red-brown solution. The resultant brown residue is extracted with 75 mL toluene. Addition of 150 mL diethyl ether, and then cooling, affords 5.0 g brown solid. The solid is recrystallized from toluene with hexane addition and then cooling, affording 3.67 g golden-orange crystalline blocks. A second crop of 1.25 g is obtained (combined yield 65%).

Anal. Calcd. $C_{17}H_{41}P_3ClIr$ $[IrCl(C_8H_{14})(PMe_3)_3]$: C, 36.07; H, 7.30. Found: C, 35.91; H, 7.12.

A pressure bottle containing an unstirred golden solution of 3.0 g $IrCl(C_8H_{14})$-$(PMe_3)_3$ (5.3 mmole) in 80 mL toluene is evacuated and then pressured to 20 psig with CO_2. After 4 days, the resultant mixture is filtered, hexane washed, and then dried, affording 2.50 g light cream-colored crystals (87% yield).

Method 2. The same compound may be made by a very different route: To a solution of 2.0 g (2.2 mmole) $Ir_2Cl_2(C_8H_{14})_4{}^5$ in 150 mL toluene, 1.40 g (18 mmole) trimethylphosphine in 10 mL toluene is added dropwise. After 3 days the mixture is filtered, and the resultant solid is washed with hexane. Drying to <0.1 micron affords 2.0 g (85% yield) of dark-orange solid, mp (vac.) >207° (dec.).

Anal. Calcd. for $C_{12}H_{36}P_4ClIr$ ([Ir(PMe$_3$)$_4$] Cl): C, 27.09; H, 6.82. Found: C, 27.45; H, 6.70. The reactivity of this complex with polar organic solvents that dissolve it hampers purification by recrystallization.

A 10 mL Hastelloy C shaker tube (or other appropriate high pressure reactor) containing 1.0 g [Ir(PMe$_3$)$_4$] Cl, 2 mL diethyl ether, and 2 g CO_2 is agitated at 70° for 15 hours, affording a colorless mixture. Filtration of the mixture followed by a diethyl ether wash affords an off-white solid with the IR spectrum of $Ir(C_2O_4)(Cl)(PMe_3)_3$. Rapid recrystallization from acetonitrile with diethyl ether and cooling affords white crystals with the same IR spectrum.

Anal. Calcd. for $C_{11}H_{27}O_4P_3ClIr$ [Ir(C$_2$O$_4$)(Cl)(PMe$_3$)$_3$] : C, 24.29; H, 5.00; O, 11.77. Found: C, 24.63; H, 5.33; O, 10.12.

Properties

An X-ray crystal structure determination[3] has shown the configuration indicated in structure 2. NMR studies have been prevented by the short-term stability of this complex in solvents such as acetonitrile and acetone. Infrared bands at 1725(s), 1680(s), 1648(sh), 1605(m), 1290(s), 1005(m), and 790(m) cm^{-1} have been assigned to involve the bound CO_2 by isotope labeling ($^{13}CO_2$ and $C^{18}O_2$) studies. Trimethylarsine instead of PMe$_3$ affords the corresponding $Ir(C_2O_4)Cl(AsMe_3)_3$ complex, and impure preparations of the corresponding rhodium complexes result from $Rh_2Cl_2(C_2H_4)_4$ as the starting complex.

References and Notes

1. Reviews: (a) R. Eisenberg, *Adv. Catal.*, **28**, 79 (1979). (b) M. E. Vol'pin and I. S. Kolomnikov, *Organometallic Reactions*, Vol. 5, E. I. Becker and M. Tsutsui (eds.), John Wiley, New York, 1975, pp. 313-386.
2. M. Aresta, C. F. Nobile, V. G. Albano, E. Forni, and M. Manassero, *J. Chem. Soc. Chem. Comm.*, 636 (1975).
3. T. Herskovitz and L. J. Guggenberger, *J. Am. Chem. Soc.*, **98**, 1615; (1976). U.S. Patent 3,954,821.
4. T. Herskovitz, *J. Am. Chem. Soc.*, **99**, 2391 (1977). U.S. Patent 3,954,821.
5. J. L. Herde, J. C. Lambert, and C. V. Senoff, *Inorg. Synth.*, **15**, 18 (1974).
6. R. Cramer, *Inorg. Synth.*, **15**, 14 (1974).
7. The checkers (C. K. and H. D. K.) report that the cyclometallation product from Ir(dmpe)$_2$Cl adds CO_2 to form the CO_2 insertion product.[4] The [HIr(dmpe-H)(dmpe)] Cl is formed in 98% yield while the CO_2 insertion product gives a 90% yield.
8. J. C. Calabrese and T. Herskovitz, to be published.

23. TRIS(ACETONITRILE)NITROSYLBIS(TRIPHENYL-PHOSPHINE)IRIDIUM(III) BIS[HEXAFLUOROPHOSPHATE]

Submitted by MAURO GHEDINI* and GIULIANO DOLCETTI*
Checked by WILLIAM M. GRAY† and DAVID L. THORN†

$$Ir(NO)I_2(PPh_3)_2 + 2Ag[PF_6] \xrightarrow{CH_3CN}$$

$$[Ir(NO)(CH_3CN)_3(PPh_3)_2][PF_6]_2 + \cdots$$

The discovery that the nitrosyl ligand is capable of binding to transition metals in two isomeric valence forms[1] is one of the most dramatic recent developments in organometallic chemistry. Since "bent NO" donates 2 fewer electrons to the metal than the linear isomer does, linear-bent tautomerism raises the possibility of coordinative unsaturation and catalysis.[2] In fact, complexes of nitric oxide are receiving increasing attention as catalysts,[3] since they are more reactive than the corresponding carbonyls.[4]

The complexes reported below are convenient precursors of a variety of transition metal nitrosyl complexes.[5] Since all these compounds and intermediates are more or less air sensitive, operations should be carried out using standard inert atmosphere techniques.[6] All the solvents must be purified and deoxygenated prior to use.

Procedure

A Schlenk apparatus, illustrated in Fig. 1, similar to that employed in other syntheses,[7] is used for the reaction sequence. Vacuum and argon sources are provided by a double manifold to which the stopcocks of the apparatus are attached. Prepurified nitrogen can be used in place of argon. The starting material, $Ir(NO)I_2(PPh_3)_2$, is prepared as reported in the literature.[8] In a typical preparation, $Ir(NO)I_2(PPh_3)_2$ (1.40 mmole) is introduced in the Schlenk tube A equipped with a magnetic stirrer. Then the apparatus is completed with the filter B and another Schlenk tube C. The apparatus is degassed through several pump and flush cycles, then 30 mL deoxygenated, freshly distilled acetonitrile is added, either by means of a gas-tight syringe previously flushed with nitrogen or using the flexible-needle technique, through a rubber septum placed on the stopcock. (Solvents are dried and distilled under nitrogen and stored over molecular

*Department of Chemistry, University of Calabria, 87030 Arcavacata (Cosenza), Italy.
Present address: Institute of Chemistry, University of Udine, 33100 Udine, Italy.
†Central Research and Development, E. I. du Pont de Nemours and Co., Wilmington, DE 19898.

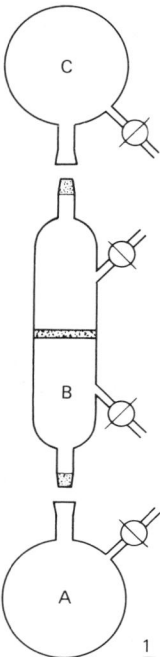

1   *Fig. 1. Schlenk apparatus.*

sieves.) Next, silver hexafluorophosphate(1-) (0.73 g; 2.9 mmoles), (Alfa Inorganic, Beverly, MA 01915) in 20 mL deoxygenated freshly distilled aceto-nitrile is added with a flexible needle. The mixture is magnetically stirred at room temperature in the dark for 12 hours until the precipitate color is that of the corresponding silver halide.

The silver iodide precipitate is removed with filtration by inverting the appar-atus. The resulting green filtrate solution is concentrated to a small volume (10-15 mL) using the pump connected to the apparatus stopcocks. Diethyl ether (40-50 mL) is then slowly added with stirring; this precipitates the appropriate nitrile complex as green crystals. Filtration can be accomplished in a new Schlenk tube or inverting again the same apparatus. The precipitate, $[Ir(NO)(CH_3CN)_3(PPh_3)_2] [PF_6]_2$, (1), is washed with diethyl ether and then dried under vacuum (yield 70%). Compound 1 can be recrystallized by addition of diethyl ether to an acetonitrile solution of the complex.

The pale-green bis(acetonitrile)nitrosyliridium complex $[Ir(NO)(CH_3CN)_2-(PPh_3)_2] [PF_6]_2$, (2), is precipitated from a dichloromethane solution of 1 by diethyl ether addition. (The checkers recommend a second filtration through a fine frit before addition of diethyl ether. They also suggest the use of 40 mL dichloromethane to dissolve 0.10 g of 1.)

NOTE: All the solvents must be absolutely anhydrous to avoid the formation of the hydroxo cation[5,8] $[Ir(NO)(OH)(PPh_3)_2]^+$.

Anal. Calcd. for **1**, $C_{42}H_{39}F_{12}IrN_4OP_4$: C, 43.49; H, 3.38; N, 4.83. Found: C, 43.26; H, 3.41; N, 4.72. Calcd. for **2**, $C_{40}H_{36}F_{12}IrN_3OP_4$: C, 42.94; H, 3.24; N, 3.75. Found: C, 43.19; H, 3.32; N, 3.65.

A recent X-ray crystal structure determination shows that in complex **1**, the acetonitrile ligand *trans* to the nitrosyl ligand has the shortest M-L distance.[9]

Properties

The complexes reported above are green, crystalline solids (mp **1**, 218-220° dec.; **2**, 190-195° dec.) that are moderately stable in air (decomposition over a period of several months is noted). The complexes are very soluble in acetonitrile and just moderately soluble in nitromethane and dichloromethane. Their solutions are not very stable in air. They are insoluble in alcohols, benzene, diethyl ether, and hexane. The complexes exhibit a nitrosyl stretching frequency at 1541 cm^{-1} (**1**) and 1684 cm^{-1} (**2**). The acetonitrile ligands have a weak combination band at 2285 cm^{-1} (**1**) and 2315 cm^{-1} (**2**) as Nujol mulls. Conductivity data in nitromethane solution [150.3 ohm^{-1} cm^2 mole^{-1} (**1**) and 154.3 ohm^{-1} cm^2 mole^{-1} (**2**)] are consistent with 2:1 electrolytes. The iridium complexes are precursors of other complexes with potential activity in homogeneous catalysis.[4,5,10]

References

1. B. A. Frenz and J. A. Ibers, *M.T.P. Int. Rev. Sci. Phys. Chem. Ser. 1*, **11**, 33 (1972).
2. J. P. Collman, G. Dolcetti, and P. Farnham, *J. Am. Chem. Soc.*, **93**, 1788 (1971).
3. G. Dolcetti and N. W. Hoffman, *Inorg. Chim. Acta*, 9, 269 (1974), and references therein. E. Zuech, W. Hughes, D. Kubicek, and E. Kettleman, *J. Am. Chem. Soc.*, **92**, 528 (1970).
4. J. P. Collman, N. Hoffman, and D. E. Morris, *J. Am. Chem. Soc.*, **91**, 5659 (1969).
5. M. Ghedini, G. Dolcetti, B. Giovannitti, and G. Denti, *Inorg. Chem.*, **16**, 1725 (1977). M. Ghedini, G. Denti, and G. Dolcetti, *Isr. J. Chem.*, **15**, 271 (1977).
6. D. F. Shriver, *The Manipulation of Air-Sensitive Compounds*, McGraw-Hill Book Co., New York, 1969, Chap. 7.
7. G. Dolcetti, M. Ghedini, and C. A. Reed, *Inorg. Synth.*, **16**, 29 (1976).
8. S. A. Robinson and M. F. Ultley, *J. Chem. Soc., A*, 1254 (1971).
9. M. Lanfranchi, A. Tiripicchio, G. Dolcetti, and M. Ghedini, *Transition Met. Chem.*, **5**, 21 (1980).
10. B. Giovannitti, M. Ghedini, G. Dolcetti, and G. Denti, *J. Organometal. Chem.*, **157**, 457 (1978).

24.   STABLE COPPER(I) CARBONYL COMPLEXES

Submitted by O. M. ABU SALAH,* M. I. BRUCE,† and C. HAMEISTER†
Checked by WM. S. DURFEE‡ and F. L. URBACH‡

Although the chemistry of copper(I) carbonyl complexes containing nitrogen donor ligands extends back far into the nineteenth century, only a few stable complexes have been isolated.[1] One of the first of these was the hydrotris(pyrazolyl)borate complex, $Cu(CO)[(pz)_3BH]$ (pz = 1-pyrazolyl, $C_3H_3N_2$), first reported in 1972,[2] the structure of which was described in 1975.[3] Other copper(I) carbonyl complexes containing substituted poly(pyrazolyl)borate ligands were subsequently obtained,[4] together with a variety of related complexes containing ligands other than CO. Other reports of solid copper(I) carbonyl complexes include those describing the unstable $Cu(CO)(O_2CCF_3)$[5] and $Cu(CO)(\eta\text{-}C_5H_5)$[6] derivatives, the five-coordinate macrocyclic ligand derivative $Cu(CO)(1bf_2)$ [$1bf_2$ = difluoro-[[3,3'-(trimethylenedinitrilo)bis(2-butanone) dioximato] (2-)] borate],[7] a binuclear $Cu^ICu^{II}(CO)$ complex containing a ligand derived from 5-methyl-2-hydroxyisophthalaldehyde and 1,3-diaminopropane,[8] and the cationic complex $[Cu(CO)(dien)]BPh_4$ [dien = diethylenetriamine, $NH(CH_2CH_2NH_2)_2$].[9]

The poly(pyrazolyl)borate ligands were developed by Trofimenko[10] and have been used by him and others as cyclopentadienyl-like ligands which are capable of forming complexes that are often more stable than the analogous η^5-cyclopentadienyl complexes. We have used this property to prepare a series of isolable, relatively stable copper(I) carbonyl complexes. The detailed syntheses of three complexes of this type, $Cu(CO)[(pz)_3BH]$, $Cu(CO)[(Me_2pz)_3BH]$ (Me_2pz = 3,5-dimethyl-1-pyrazolyl), and $Cu(CO)[(pz)_4B]$, are described below.

■ **Caution.** *These syntheses use carbon monoxide, a highly poisonous, colorless and odorless gas. The preparations should therefore be carried out in a well-ventilated fume hood.*

*Department of Chemistry, Faculty of Medicine, University of Riyadh, Riyadh, Saudi Arabia.
†Department of Physical and Inorganic Chemistry, University of Adelaide, Adelaide, South Australia 5001.
‡Department of Chemistry, Case Western Reserve University, Cleveland, OH 44106.

A. CARBONYL[HYDROTRIS(PYRAZOLATO)BORATO]COPPER(I), Cu(CO)[(pz)₃BH]

$$CuCl + KBH(pz)_3 + CO \longrightarrow Cu(CO)[(pz)_3BH] + KCl$$

Procedure

The reaction is carried out in a Schlenk tube, Fig. 1. Acetone (10 mL) is cooled in an ice bath and saturated with carbon monoxide by passing in a brisk stream of the gas for 10 minutes. Potassium hydrotris(pyrazolato)borate (252 mg, 1.0 mmole) and copper(I) chloride (99 mg, 1 mmole) are added to the acetone, and the mixture is stirred for 30 minutes while continuing to pass a slow stream of carbon monoxide. At the end of this period, the white suspension is filtered, using the Schlenk-type apparatus illustrated, Fig. 2, keeping the whole under an atmosphere of CO. The clear colorless filtrate is evaporated to dryness, and the cream-colored residue is extracted with 10 mL light petroleum (boiling range 40-60°, saturated with CO) and filtered. Evaporation of the solvent from the filtrate affords a white powder of Cu(CO)[(pz)₃BH] (156 mg, 51%).

Anal. Calcd. for $C_{10}H_{10}BCuN_6O$: C, 39.4; H, 3.3; N, 27.55. Found: C, 39.7; H, 3.6; N, 27.4.

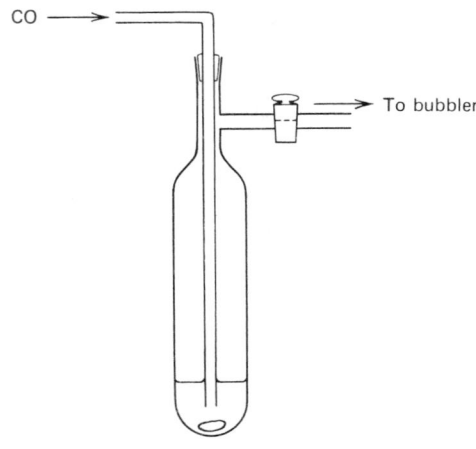

CO ⟶

To bubbler

Fig. 1. Schlenk reaction apparatus.

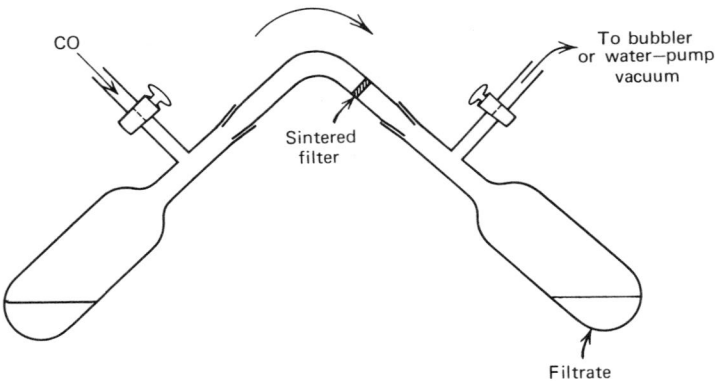

Fig. 2. Schlenk filtration apparatus

B. CARBONYL[TRIS(3,5-DIMETHYLPYRAZOLATO)HYDROBORATO]-COPPER(I), Cu(CO)[(Me₂pz)₃BH]

$$CuCl + KBH(Me_2pz)_3 + CO \longrightarrow Cu(CO)[(Me_2pz)_3BH] + KCl$$

Procedure

A two-neck, 100-mL, round-bottom flask is equipped with a sintered Dreschel head. Acetone (30 mL) is placed in the flask and saturated with CO by passing in a brisk stream of the gas for 5 minutes. Copper(I) chloride (99 mg, 1 mmole) is added and, to the stirred suspension, potassium tris(3,5-dimethylpyrazolato)-hydroborate (336 mg, 1 mmole) is added.[11] The reaction mixture is stirred for 1½ hours at room temperature while passing in a brisk stream of CO. After this time, the remaining acetone is removed from the white suspension, using a water-pump vacuum and warming gently in a water bath held at 30-40°, to give a white residue.

The residue is extracted twice with freshly distilled CO-saturated ether (100 mL), while the reaction vessel is kept under a CO atmosphere. Filtration of the extract and collection of the filtrate is performed using a conventional Schlenk-type apparatus, which is firstly evacuated (water pump) and then kept under a CO atmosphere. Slow filtration of the extract leaves a white residue and gives a clear colorless filtrate. The volume of the filtrate is reduced to ca. 20 mL by evaporating in vacuo with gentle warming on a water bath (30-35°). A white powder separates and is collected by filtration in a Schlenk apparatus flushed with CO; a further crop is obtained by reducing the volume of filtrate to ca. 10 mL. The product is washed with cold CO-saturated light petroleum (5 mL) and dried in vacuo. Yield 190 mg (50%).

Anal. Calcd. for $C_{16}H_{22}BCuN_6O$: C, 49.5; H, 5.65; N, 21.6. Found: C, 49.0; H, 5.85; N, 21.85.

C. CARBONYL[TETRAKIS(PYRAZOLATO)BORATO]COPPER(I), $Cu(CO)[(pz)_4B]$

$$CuCl + KB(pz)_4 + CO \longrightarrow Cu(CO)[(pz)_4B] + KCl$$

Procedure

Potassium tetrakis(pyrazolato)borate (318 mg, 1 mmole)[11] is added to a suspension of copper(I) chloride (99 mg, 1 mmole) in CO-saturated acetone (30 mL) contained in a 100-mL, two-neck round-bottom flask equipped with a sintered Dreschel head. The reaction mixture is stirred for 2 hours at room temperature while a brisk stream of CO is passed through it. After this time, more CO-saturated acetone (90 mL) is added to the white suspension and the mixture heated to boiling under CO. Filtration and collection of the filtrate is performed in a conventional Schlenk apparatus which has been previously flushed with CO. Failure to maintain a CO flush leads to a blue tinted solution and product. Fast filtration using a water-pump vacuum leaves a white residue and a clear, colorless filtrate. The residue is extracted again with CO-saturated acetone and filtered as before. The volume of the combined filtrates is reduced to ca. 10 mL in vacuo (water pump) while warming gently on a water bath (40-50°). White crystals separate. These are filtered off using the Schlenk apparatus, washed with CO-saturated light petroleum, and dried in vacuo. (It is important to dry the solid thoroughly. If this is not done, the solid turns blue immediately in air.) Yield 167 mg (46%).

Anal. Calcd. for $C_{13}H_{12}BCuN_8O$: C, 42.05; H, 3.25; N, 30.2. Found: C, 41.8; H, 3.4; N, 30.0.

Properties

(a) $Cu(CO)[(pz)_3BH]$. This complex forms white crystals mp 164-166°. The solid is stable in a CO atmosphere and slowly turns very pale blue in air. In solution, oxidation to a blue solution is quite rapid. The infrared spectrum contains a single sharp ν(CO) band at 2079 cm^{-1} and a broader ν(BH) absorption at 2465 cm^{-1}. The ^{1}H nmr spectrum contains resonances at τ 2.30d and 3.83t. Thermal decomposition results in loss of the carbonyl ligand, followed by disproportionation to give blue Cu[(pz)$_3$BH]$_2$ and copper metal. The carbonyl group is also displaced by a variety of ligands, L, to give the complexes Cu(L)[(pz)$_3$BH], most of which are stable in air. Examples with L = PR$_3$, P(OR)$_3$, AsR$_3$, SbR$_3$, and CNR have been described.[2]

(b) **Cu(CO)[(Me$_2$pz)$_3$BH]**. The dimethylpyrazolato complex forms white crystals which turn brown at 187-189° (sealed tube). The solid appears to be stable in a CO atmosphere but turns blue in air over a day. The infrared spectrum contains a single sharp ν(CO) absorption at 2063 cm^{-1} and a ν(BH) band at 2520 cm^{-1}.

(c) **Cu(CO)[(pz)$_4$B]**. The tetrakis(pyrazolato)borate complex forms white crystals which turn pale brown at 135-138° (sealed tube). This complex is the least stable of the three carbonyls described, and solutions of the carbonyl turn blue immediately on contact with air. The infrared spectrum contains a single sharp ν(CO) band at 2083 cm^{-1}, and the ^{1}H nmr spectrum contains only three signals for the pyrazolyl protons, at τ 2.28s, 2.34s, and 3.84s, suggesting that the B(pz)$_4$ group is fluxional, undergoing a random dissociation-association "tumbling" process. This is confirmed by low-temperature studies of the spectra of related complexes containing tertiary phosphine ligands.[12]

References

1. M. I. Bruce, *J. Organometal. Chem.*, **44**, 209 (1972). F. H. Jardine, *Adv. Inorg. Chem. Radiochem.*, **17**, 115 (1975).
2. M. I. Bruce and A. P. P. Ostazewski, *J. Chem. Soc. Chem. Comm.*, 1124 (1972). M. I. Bruce and A. P. P. Ostazewski, *J. Chem. Soc., Dalton*, 2433 (1973).
3. M. R. Churchill, B. G. DeBoer, F. J. Rotella, O. M. Abu Salah, and M. I. Bruce, *Inorg. Chem.*, **14**, 2051 (1975).
4. O. M. Abu Salah, M. I. Bruce, and J. D. Walsh, *Aust. J. Chem.*, **32**, 1209 (1979).
5. A. F. Scott, L. L. Wilkening, and B. Rubin, *Inorg. Chem.*, **8**, 2533 (1969).
6. F. A. Cotton and T. J. Marks, *J. Am. Chem. Soc.*, **92**, 5114 (1970).
7. R. R. Gagné, *J. Am. Chem. Soc.*, **98**, 6709 (1976). R. R. Gagné, J. L. Allison, R. S. Gall, and C. A. Koval, *J. Am. Chem. Soc.*, **99**, 7170 (1977).
8. R. R. Gagné, C. A. Koval, and T. J. Smith, *J. Am. Chem. Soc.*, **99**, 8367 (1977).
9. M. Pasquali, F. Marchetti, and C. Floriani, *Inorg. Chem.*, **17**, 1684 (1978).
10. S. Trofimenko, *Acc. Chem. Res.*, **4**, 17 (1971). *Chem. Rev.*, **72**, 497 (1972).
11. S. Trofimenko, *Inorg. Synth.*, **12**, 99 (1970).
12. O. M. Abu Salah and M. I. Bruce, *J. Organometal. Chem.*, **87**, C15 (1975).

25.   {TRIS[μ-[(1,2-CYCLOHEXANEDIONE DIOXIMATO)-$O{:}O'$]DIPHENYLDIBORATO(2-)]-$N,N',N'',N''',N'''',N'''''$}-IRON(II)

Submitted by J. N. JOHNSON* and N. J. ROSE*
Checked by J. W. BRANCH† and V. L. GOEDKEN†

$$FeCl_2 \cdot 4H_2O + 3H_2nox + 3\emptyset B(OH)_2 \xrightarrow{\ H_2O\ } [Fe\{(nox)_3(B\emptyset)_2\}] +$$
$$2H^+ + 10H_2O + 2Cl^-$$

Clathrochelates of the tris(glyoxime) class are readily made in high yield when BF_3 or $B(OH)_3$ is used as the capping agent.[1] Alternately, the use of $B(OH)_3$ in conjunction with various alcohols (ROH) leads to a variety of cages which are capped with the BOR moiety.[1] The following description of a tris(glyoxime) cage demonstrates that an even wider diversity of structural type can be envisioned and readily attained by using $R'B(OH)_2$ as the capping reagent. It is anticipated that such variation will be important to investigators wishing to take advantage of the intrinsic properties of the cages (e.g., high extinction coefficients, thermal stability, etc.) while (1) modifying their physical properties such as solubility or (2) incorporating certain functional groups in the R' of $R'B(OH)_2$ such that the resulting cages will in turn be starting materials for other reactions (e.g., polymerization).

The nomenclature used to represent the clathrochelate is from Reference 1. The chemicals employed in the synthesis are commercially available (reagent grade wherever possible) and are used without further purification. Elemental analyses were performed by Chemalytics, Inc., Tempe, AZ.

Procedure

Iron chloride ($FeCl_2 \cdot 4H_2O$, 5mmole, 0.993 g) is dissolved in 60 mL distilled water contained in a 125-mL Erlenmeyer flask. A stirring bar is added and the reaction mixture is *stirred vigorously* throughout the procedure. Cyclohexanedione dioxime (nioxime) (15mmole, 2.13 g) is added, as well as dihydroxyphenylborane (phenylboronic acid) (10mmole, 1.22 g) dissolved in 5 mL ethanol. The reaction mixture is stirred for about 30 minutes, and then the orange powdery product which has formed is collected on a 150-mL, medium sintered-glass funnel with suction filtration. The crude product is first washed with two 20-mL portions of acetone and then two 20-mL portions of diethyl

*Department of Chemistry, University of Washington, Seattle, WA 98195.
†Department of Chemistry, Florida State University, Tallahassee, FA 32306.

ether. It is dried by sucking air through it for 30 minutes. The yield is essentially quantitative.

Recrystallization proceeds as follows. A mixture of the crude product (1 g) and 150 mL dichloromethane is heated to 35° for 10 minutes with stirring. The solution is filtered by gravity (Whatman #1 paper). Ethanol (~250 mL) is added to the filtrate *slowly* and just until precipitation begins to occur (NOTE: Addition of excess ethanol causes the product to precipitate rather than separate in crystalline form.) The solution is then allowed to cool *slowly* to ~2° and is kept at that temperature for a few days. This is accomplished by first placing the flask containing the hot filtrate inside an insulated container such as a Styrofoam bucket (which is at room temperature) for several hours and then placing the entire assembly (insulator and flask) inside a refrigerator held at ~2°. The crystals are collected on a medium sintered-glass frit with suction filtration and washed with two 20-mL portions of acetone. They are further dried by placing them over P_4O_{10} for 24 hours at approximately 1 torr. Approximate yield: 0.5 grams (50%).

Anal. Calcd. for $C_{30}H_{34}N_6O_6B_2Fe$: C, 55.24; H, 5.26; N, 12.89. Found: C, 55.47; H, 5.36; N, 12.91.

Properties

The product has appreciable solubility in dichloromethane and more limited solubility in both benzene and pyridine. The Mossbauer spectrum and electronic spectrum of this clathrochelate are similar to those of other iron tri(glyoxime) cages.[2] The compounds show an intense absorption band at 22,200 cm^{-1}(ϵ = 17,900). Several dominant bands in the infrared spectrum along with their proposed assignments are: 1589 cm^{-1} (C=N stretch); 1495 cm^{-1} (C−C and C=N stretch); 1234 and 1064 cm^{-1} (N−O stretch), and 1209 cm^{-1} (B−O stretch).[3] In dichloromethane-d_2, the pmr spectrum consists of two unresolved triplets (relative area = 24) at 1.78 and 2.91 ppm downfield from internal TMS and two multiplets (relative area 10.3) at 7.3 and 7.7 ppm downfield.[3] The two multiplets are strikingly similar to that observed in the spectrum of phenyl-

$[Fe\{(nox)_3 (B\phi)_2\}]$

boronic acid itself. In benzene the molecular weight is found to be 631 (calcd. 652) via vapor phase osmometry.[3]

References

1. S. C. Jackels, J. Zektzer, and N. J. Rose, *Inorg. Synth.*, **17**, 139 (1977).
2. S. C. Jackels and N. J. Rose, *Inorg. Chem.*, **12**, 1232 (1973).
3. J. N. Johnson, thesis, University of Washington, 1976.

26. TERNARY COMPLEXES OF COPPER(II)

Submitted by W. L. KWIK* and K. P. ANG*
Checked by M. WINKLER,† D. SPIRA,† and E. I. SOLOMON†

There has been considerable interest in the chemistry of ternary complexes of copper(II) containing a bidentate aromatic nitrogen base such as 1,10-phenanthroline (phen) or 2,2'-bipyridine (bpy) and a bidentate oxygen donor ligand or an amino acid,[1-6] as some of these could possibly serve as models for enzyme-metal ion-substrate complexes. Two procedures are described below for the convenient, high-yield preparation of two such complexes.

A. [MALONATO(2–)](1,10-PHENANTHROLINE)COPPER(II) DIHYDRATE

$$CuCl_2 \cdot 2H_2O + phen \longrightarrow Cu(phen)Cl_2 + 2H_2O$$

$$Cu(phen)Cl_2 + mal \longrightarrow Cu(phen)[mal(2-)] + 2HCl$$

Procedure

A sample of 1,10-phenanthroline monohydrate (3.96 g, 20 mmole) in 20 mL ethanol is added to a solution of copper(II) chloride dihydrate (3.40 g, 20 mmole) in 40 mL ethanol. A suspension of a green-blue solid develops immediately. A solution of malonic acid (2.08 g, 20 mmole) in 50 mL ethanol is then slowly introduced into this suspension, which is being stirred magnetically. This

*Department of Chemistry, National University of Singapore, Kent Ridge, Singapore, 0511.
†Dept. of Chemistry, Massachusetts Institute of Technology, Cambridge, MA 02139.

is followed by a gradual addition of 1.0 M ammonium hydroxide solution until a blue precipitate is formed. This solid is separated through filtration and washed with several 10-mL portions of distilled water as well as ethanol. The product is then recrystallized from water/methanol (3:2) mixture, and the bright blue crystals thus obtained are air dried and finally dried in vacuo. Yield is 5.0-5.4 g [70-75% based on $CuCl_2 \cdot 2H_2O$].

Anal. Calcd. for $Cu(phen)(mal) \cdot 2H_2O$; Cu, 16.60; C, 47.20; H, 3.70; N, 7.30. Found: Cu, 16.80; C, 46.90; H, 3.80; N, 7.20.

B. (1,10-PHENANTHROLINE)[SERINATO (1−)]COPPER(II) SULFATE

$$CuSO_4 \cdot 5H_2O + phen \longrightarrow Cu(phen)(SO_4) + 5H_2O$$

$$2Cu(phen)(SO_4) + 2ser \longrightarrow 2\{Cu(phen)[ser\ (1-)]\}_2(SO_4) + H_2SO_4$$

Procedure

To a solution of copper(II) sulfate pentahydrate (5.0 g, 20 mmole) in 20 mL distilled water, a solution of 1,10-phenanthroline monohydrate (6.3 g, 32 mmole) in 20 mL ethanol is slowly added. As the resulting blue suspension is being stirred vigorously in a 250-mL round-bottom flask, a sample of L-serine (3.36 g, 32 mmole) in 10 mL 0.1 M hydrochloric acid is added. Approximately 150 mL 1.0 M ammonium hydroxide solution is then gradually introduced into the mixture till a clear blue solution ensues. This solution is gently refluxed for about ½ hour before its volume is reduced to approximately half by gentle heating over a hot plate. On cooling in ice, the resultant solution yields a blue solid, leaving behind a deep green filtrate. The solid is separated by filtration, washed repeatedly with portions of 10 mL cold, distilled water and ethanol, air dried, and finally dried in vacuo. Yield is 4.2-4.5 g (50-55% based on $CuSO_4 \cdot 5H_2O$).

Anal. Calcd. for $[Cu(phen)(ser)]_2 (SO_4) \cdot 2H_2O$: Cu, 15.90; C, 43.50; H, 3.90; N, 10.10; Found: Cu, 16.60; C, 43.40; H, 4.10; N, 9.70.

Properties

These copper(II) ternary complexes are generally green-blue solids, stable in air, and fairly soluble in water, methanol, and ethanol. Furthermore, those with dicarboxylates and amino acid residues exhibit a characteristic, strong infrared absorption band around 1600 cm^{-1} due to the coordinated carboxylate group. The infrared spectra from 4000-200 cm^{-1} as well as the electronic spectra of these complexes have been recorded and assignments made.[5,6]

References

1. G. A. L'Herureux and A. E. Martell, *J. Inorg. Nucl. Chem.*, **28**, 481 (1966).
2. H. Sigel and D. B. McCormick, *Acc. Chem. Res.*, **3**, 201 (1970).
3. P. R. Huber, R. Griesser, and H. Sigel, *Inorg. Chem.*, **10**, 945 (1971).
4. F. A. Walker, H. Sigel, and D. B. McCormick, *Inorg. Chem.*, **11**, 2756 (1972).
5. W. L. Kwik and K. P. Ang, *Aust. J. Chem.*, **31**, 459 (1978).
6. W. L. Kwik, K. P. Ang, and G. Chen, *J. Inorg. Nucl. Chem.*, **42**, 303 (1980).

27. PREPARATIONS OF *trans*-DIOXOTETRAKIS(PYRIDINE)-RHENIUM(V) CHLORIDE, [ReO$_2$py$_4$] Cl, AND PERCHLOROATE, [ReO$_2$py$_4$] ClO$_4$

Submitted by M. C. CHAKRAVORTI*
Checked by R. A. ANDERSEN†

■ **Caution.** *All perchlorate salts are potentially dangerously explosive and should be handled with extreme care.*

Cationic complexes of rhenium are very few. One of the thoroughly studied complexes is [ReO$_2$py$_4$]Cl. The preparation of the analogous 1,2-ethanediamine complex has earlier appeared[1] in *Inorganic Syntheses*. The pyridine complex has been prepared in several ways including (a) reaction of K$_2$[ReOCl$_5$][2] or K$_2$[ReCl$_6$][3,4] with aqueous pyridine, (b) reaction of [ReOCl$_3$(PPh$_3$)$_2$] and pyridine in ethanol,[5] (c) boiling the brown product obtained by the reaction between potassium perrhenate (KReO$_4$), potassium iodide, and hydrochloric acid with aqueous pyridine.[6] All these chloro complexes, namely, K$_2$[ReOCl$_5$], K$_2$[ReCl$_6$], and [ReOCl$_3$(PPh$_3$)$_2$], from which the pyridine complex has been obtained are prepared from KReO$_4$ or HReO$_4$. It is therefore of interest to develop a method of preparation of [ReO$_2$py$_4$]Cl starting directly from KReO$_4$, thus avoiding all the intermediate stages of preparation, so that the complete preparation becomes easier, less time consuming, and the yield (based on the amount of KReO$_4$) is higher.

A study of the different methods of the preparation of [ReO$_2$py$_4$]Cl shows that the same quinquevalent rhenium compound is obtained starting from Re(IV) or Re(V) chloro complex. It is known[7] that by boiling a mixture of potassium perrhenate, potassium iodide, and hydrochloric acid, a brown product is obtained which contains rhenium mostly in the 5+ oxidation state, with a little of Re(IV). Isolation of pure K$_2$[ReCl$_6$] or K$_2$[ReOCl$_5$] from the brown

*Department of Chemistry, Indian Institute of Technology, Kharagpur 721 302, India.
†Department of Chemistry, University of California, Berkeley, CA 94720.

residue is time consuming. It has been shown[6] that $[ReO_2py_4]Cl$ can be directly obtained by boiling the brown residue with aqueous pyridine. The procedure described here is based on this. The compound has been variously formulated as anhydrous, monohydrate, and dihydrate. In fact, purification of the compound is difficult; the compound is weakly hygroscopic. It loses pyridine slowly on standing and becomes difficult to dissolve. The corresponding perchlorate salt, $[ReO_2py_4]ClO_4$, can however be prepared in pure state and is stable.

Procedure

$$KReO_4 + 3KI + 8HCl \longrightarrow K_2[ReCl_6] + \frac{3}{2}I_2 \uparrow + 4H_2O + 2KCl$$

$$2K_2[ReCl_6] + 14py + \tfrac{1}{2}O_2 + 7H_2O \longrightarrow 2[ReO_2py_4]Cl \cdot 2H_2O$$
$$+ 6(pyH)Cl + 4KCl$$

An intimate mixture of 4.0 g (14mmole) of potassium perrhenate (the checker started with perrhenic acid[10]) and 5.8 g (35mmole) potassium iodide is prepared by grinding in an agate mortar. It is transferred to a porcelain dish and gently boiled with concentrated hydrochloric acid (5 mL). Most of the liberated iodine volatilizes away. The mixture is then dried on a steam bath. More acid (5 mL) is added and again evaporated. The process of evaporation with hydrochloric acid is repeated 4 times to make the mixture free from iodine. The mass is completely dried on a hot plate and allowed to cool. It is then stirred with a mixture of 15 mL pyridine (which is distilled earlier) and 40 mL water and kept overnight. The mixture is boiled under reflux for 3 hours during which time an intense red-colored solution results. This is filtered warm and the solution kept in a desiccator over sulfuric acid.

After 2 days, red crystals separate, are filtered under suction and washed twice with a small volume (2 mL) of water. The crystals are powdered and extracted with ethanol. The extract is reduced to a very small volume (almost dry) in air or in a desiccator over fused calcium chloride and filtered under suction. The product, $[ReO_2py_4]Cl \cdot 2H_2O$, is dried in air. Yields are typically 5.0 g (60% based on the amount of $KReO_4$). (The checker obtained a yield of 34%.) The anhydrous compound is obtained by drying over P_4O_{10}. The compound is of moderate purity and can be used for ordinary purposes. It can be crystallized by slow evaporation in air of an aqueous ethanolic solution (50:50) containing traces of pyridine.

For the preparation of the perchlorate, $[ReO_2py_4]ClO_4$, a concentrated solution of sodium perchlorate (2.0 g, 16.5mmole) is added to a concentrated solution of air-dried $[ReO_2py_4]Cl \cdot 2H_2O$ (2.0 g, 3.3mmole) with stirring whereupon an orange precipitate is obtained. (See cautionary remark above.) After ½ hour, the precipitate is filtered under suction and thoroughly washed with water and then twice with a small volume (0.5 mL) of ethanol. The compound is dried in a desiccator over sulfuric acid. The yield is 90%. The compound can be crystallized as needles by slow evaporation of ethanol or acetone solution.

Anal. Calcd. for $[ReO_2py_4]ClO_4$: Re, 29.40; Cl, 5.61; N, 8.84. Found: Re, 29.6; Cl, 5.60; N, 8.86.

For the analysis of rhenium, the sample is fused with sodium peroxide and rhenium precipitated from hydrochloric acid solution as Re_2S_7 by hydrogen sulfide. The precipitate is dissolved in aqueous sodium peroxide; and from dilute sulfuric acid solution, rhenium is determined by electrodeposition.[8] Chlorine is determined gravimetrically by fusion of the sample with a sodium carbonate/ sodium nitrate mixture (95:5) in a platinum crucible and then precipitating as silver chloride.

Properties

Dioxotetra(pyridine)rhenium(V) perchlorate is an orange or orange-yellow crystalline substance. It is very slightly soluble in water but soluble in acetone and ethanol. It does not lose pyridine at room temperature, but decomposes above 100°. On acidification it gives a purple-red color[4] due to the protonation of $[ReO_2py_4]^+$ to $[ReO(OH)py_4]^{2+}$. It is very weakly paramagnetic (μ_{eff} = 0.51 BM).[6] In the infrared spectrum, it gives a strong band at 823 cm^{-1}, indicating a trans dioxo (O=Re=O) structure.[4] (The trans structure in $[ReO_2py_4]$-Cl·2H$_2$O has been confirmed by X-ray diffraction.[9]) The absorption spectrum[4] in aqueous solution gives bands at 331 and 445 nm.

References

1. R. K. Murmann, *Inorg. Synth.*, 8, 173 (1966).
2. V. V. Lebedinskii and B. N. Ivanov Emin, *Russ. J. Inorg. Chem.*, 4, 794 (1959).
3. B. Sur and D. Sen, *Sci. Culture* (Calcutta), 26, 85 (1960).
4. J. H. Beard, J. Casey, and R. K. Murmann, *Inorg. Chem.*, 4, 797 (1965).
5. N. P. Johnson, C. J. L. Lock, and G. Wilkinson, *J. Chem. Soc.*, 1054 (1964).
6. M. C. Chakravorti, *J. Ind. Chem. Soc.*, 47, 827 (1970).
7. L. C. Hurd and V. A. Reinders, *Inorg. Synth.*, 1, 178 (1939).
8. B. K. Sen and P. Bandyopadhyay, *J. Ind. Chem. Soc.*, 40, 813 (1963).
9. C. Calvo, N. Krishnamachari, and C. J. L. Lock, *J. Cryst. Molec. Struct.*, 1, 161 (1971).
10. N. P. Johnson, C. J. L. Lock, and G. Wilkinson, *Inorg. Synth.*, 9, 145 (1967).

28. DIBROMO-, AQUABROMO- AND AQUACHLOROBIS(1,2-ETHANEDIAMINE)COBALT(III) COMPLEXES

Submitted by W. G. JACKSON* and C. M. BEGBIE*
Checked by A. C. DASH† and G. M. HARRIS†

A large majority of diacidobis(1,2-ethanediamine)cobalt(III) complexes can be synthesized from the (carbonato)bis(1,2-ethanediamine)cobalt(III) ion. This is now readily available in high yield and purity.[1] *cis-* and *trans-*Dichlorobis(1,2-ethanediamine)cobalt(III) chlorides are obtained from the carbonato complex with hydrochloric acid using appropriate conditions. These isomeric dichloro salts are widely used starting materials for further syntheses. In this respect, the analogous dibromo complexes can be more convenient starting materials because bromide is substituted from the cobalt center more readily than is chloride ion. The milder conditions minimize the production of cobalt(II), a common side reaction.

Werner prepared *cis-* and *trans-*dibromobis(1,2-ethanediamine)cobalt(III) bromide by reacting the carbonato bromide with concentrated hydrobromic acid.[2] In the cold, the cis isomer resulted; the trans isomer was obtained by heating the mixture. The reaction proceeds via *cis-*aquabromobis(1,2-ethanediamine)cobalt(III) dibromide which can be isolated by this route. In our hands, the preparations, as originally described, are not reproducible. The reaction times and yields are variable, depending too critically on the conditions for one not intuitively skilled in the art to consistently obtain the desired cis or trans form. However, the syntheses described here for the cis and trans dibromo complexes and based on this reaction are reproducible if one adheres closely to the procedures.

Other preparative methods have been reported.[2,3] *cis-*Dibromobis(1,2-ethanediamine)cobalt(III) bromide is obtained in reasonable yield by evaporating a neutral aqueous solution of the trans isomer to dryness. The reaction between *cis-*diazidobis(1,2-ethanediamine)cobalt(III) perchlorate and hydrobromic acid proceeds rapidly and quantitatively at ambient temperature, largely with retention of the cis configuration to yield the dibromo bromide.[4,5] The diazido starting complex[6,7] can be made within 1 hour in greater than 90% yield.

*Chemistry Department, University of New South Wales, Faculty of Military Studies, Royal Military College, Duntroon, Canberra, A.C.T., Australia, 2600.
†Department of Chemistry, State University of New York at Buffalo, Buffalo, NY 14214.

However, this latter preparation is hazardous,[5,6] and the former, although inferior, is to be preferred as a source of the cis dibromo bromide. Details are given here.

In addition to the cis and trans dibromo complexes, the syntheses of *cis*-aquabromobis(1,2-ethanediamine)cobalt(III) dibromide monohydrate and *trans*-aquabromobis(1,2-ethanediamine)cobalt(III) dithionate monohydrate are described. It is of historical interest that of the hundreds of cis/trans isomeric pairs of diacidobis(1,2-ethanediamine)cobalt(III) complexes known to Werner,[3] the trans isomers in the classical aquahalo series eluded him. The procedures described here for their isolation are a modification of much later work[5,8] in which the more soluble trans complexes were obtained pure as their sulfate salts by fractional crystallization of cis/trans mixtures. The syntheses of *cis*-$[Co(en)_2(OH_2)Cl]SO_4 \cdot 2H_2O$[1,9]   and   *cis*-$[Co(en)_2(OH_2)Cl]Br_2 \cdot H_2O$[9]   are described in previous volumes of this series. The preparation of *trans*-aquachlorobis(1,2-ethanediamine)cobalt(III) dithionate monohydrate is given here to complete the series.

With the exception of the preparation in Section 28-C (4-6 hours), each synthesis requires about 1 hour. The starting complexes for all the preparations can be obtained in 1-2 days.

A. CARBONATOBIS(1,2-ETHANEDIAMINE)COBALT(III) BROMIDE[†]

$$[Co(en)_2(CO_3)]Cl + NaBr \xrightarrow{H_2O} [Co(en)_2(CO_3)]Br + NaCl$$

Procedure

Once recrystallized (from H_2O/methanol),[1] $[Co(en)_2(CO_3)]Cl$ (50.0g) is dissolved in hot water (250 mL, 85°) and quickly filtered into a solution of sodium bromide (90 g) in water (100 mL, 20°). The mixture is shaken regularly while rapidly cooled in ice to facilitate the separation of fine, deep-pink crystals of $[Co(en)_2(CO_3)]Br$. After 1 hour at 5°, the product is filtered, washed with methanol (3 × 50 mL) followed by diethyl ether (3 × 50 mL), and dried in air. Yield 52.5 g (90%).

B. *trans*-DIBROMOBIS(1,2-ETHANEDIAMINE)COBALT(III) BROMIDE

$$[Co(en)_2(CO_3)]Br + 2HBr \longrightarrow \textit{trans-}[Co(en)_2Br_2]Br + H_2O + CO_2\uparrow$$

[†]en = 1,2-ethanediamine.

In the following, it is essential to have the hydrobromic acid hot *before* commencing the addition of the carbonato bromide. This avoids the formation of any cis product. The use of carbonato chloride in lieu of the bromide leads to a product which consistently contains a 5% chloride impurity.[4] This contaminant, *trans*-[Co(en)$_2$Br Cl] Br, is not removed by recrystallization.

Procedure

Fresh concentrated hydrobromic acid (300 mL, bromine free, AnalaR, 48% w/w; specific gravity 1.48) is heated on a hot plate to 85°. While well magnetically stirred, [Co(en)$_2$CO$_3$] Br (64 g, preparation A) is added in approximately 5-g portions over 5 minutes; CO$_2$ is immediately liberated. The initially deep violet-brown solution quickly becomes green. After 30 minutes at 85° when much of the product has crystallized as gleaming lime-green plates, the mixture is cooled to 0° in an ice/salt bath and allowed to stand for 30 minutes at this temperature. The crystals of the unstable hydrobromide salt, *trans*-[Co(en)$_2$Br$_2$] Br·2H$_2$O·HBr, crumble to a lime-green power in air, and are washed with ice-cold ethanol (3 × 100 mL) followed by diethyl ether and then dried in air. Yield 81 g (96%).

Anal. Calcd. for *trans*-[Co(C$_4$H$_{16}$N$_4$)Br$_2$] Br: C, 11.47; H, 3.85; N, 13.38; Br, 57.24. Found: C, 11.50; H, 3.91; N, 13.50; Br, 57.32%. The product is analytically and isomerically pure. A crystalline sample may be obtained from a hot saturated methanol solution by careful addition of diethyl ether and cooling.

C. *cis*-DIBROMOBIS(1,2-ETHANEDIAMINE)COBALT(III) BROMIDE MONOHYDRATE

$$1. \quad trans\text{-}[Co(en)_2Br_2] \, Br \xrightarrow{\quad H_2O \quad} cis\text{-}[Co(en)_2Br_2] \, Br\cdot H_2O$$

Procedure

trans-[Co(en)$_2$Br$_2$] Br (15.0 g) is dissolved in water (50 mL) by warming, and the solution is taken to near dryness in an evaporating basin on a steam bath. The grey-black crust of *cis*-[Co(en)$_2$Br$_2$] Br is broken up regularly in the final stages. The cooled mixture is filtered and washed with ice water (2 × 50 mL). This filtrate and the washings are taken to dryness to give a second crop of the crude product. The combined crude solids are transferred to a beaker using ethanol and stirred well in this solvent (200 mL) to break up the lumps. Most of the residual trans isomer is leached out by this process. The collected black-grey product is washed with ethanol (3 × 20 mL) and then diethyl ether (3 × 20 mL) and dried in air. The solid (13.6 g, 87% yield) is essentially pure *cis*-[Co(en)$_2$Br$_2$] Br and is suitable for most further synthetic work. It may be recrystallized as follows to remove small amounts of *trans*-[Co(en)$_2$Br$_2$] Br and *cis*-[Co(en)$_2$(OH$_2$)Br] Br$_2$·H$_2$O. A finely ground 20.0-g sample is shaken vigorously with water (500 mL, 20°) for 30 seconds, and the suspension is filtered immediately into a mixture of concentrated hydrobromic acid (100 mL) and ice (100 g) chilled in an ice/salt bath. Two more such extractions of the residue on the filter dissolve all but a trace, which is discarded. A further portion (100 mL) of acid is added to the combined filtrates (approximately 1700 mL) in which much of the product has already crystallized. After 1 hour at 0°, the small gleaming charcoal-grey plates of pure *cis*-[Co(en)$_2$Br$_2$] Br·H$_2$O are collected and washed and dried as above. Yield 12.5 g; 62.5% recovery; overall yield 54%.

Anal. Calcd. for *cis*-[Co(C$_4$H$_{16}$N$_4$)Br$_2$] Br·H$_2$O: C, 11.00; H, 4.15; N, 12.83; Br, 54.88. Found: C, 11.20; H, 4.30; N, 12.65; Br, 54.62.

$$\text{2.}\quad [Co(en)_2(CO_3)] Br + 2HBr \longrightarrow cis\text{-}[Co(en)_2Br_2] Br·H_2O + CO_2\uparrow$$

Procedure

This method is quicker, but the yield is lower. The sequence and timing of the addition of the reagents are important.

To a well-stirred portion of concentrated hydrobromic acid (100 mL, 20°) is added [Co(en)$_2$(CO$_3$)] Br (15.0 g) in 3-g lots within 1 minute. The pink-red carbonato complex dissolves to a deep-brown to olive-green solution with the vigorous evolution of CO$_2$. The reaction is mildly exothermic. After a few minutes, olive-green crystals appear and the now warm mixture is stirred for a total of 30 minutes. The grey to olive-green crystals are collected, washed with concentrated hydrobromic acid (3 × 20 mL), cold ethanol (3 × 20 mL), and finally diethyl ether (3 × 20 mL). The air-dried product (18.9 g) is a mixture of *cis*- and *trans*-[Co(en)$_2$Br$_2$] Br. A single recrystallization from water as described in preparation 1 in this section gives pure *cis*-[Co(en)$_2$Br$_2$] Br·H$_2$O; the trans isomer does not reprecipitate under these conditions. Yield 5.4 g, overall yield 26%.

Anal. Found: C, 11.25; H, 4.35; N, 12.74; Br, 54.72%.

Properties

The cis and trans dibromo complexes resemble the dichloro analogs, the properties of which are recorded in References 1, 3, and 9. In solution, *cis*-[Co(en)$_2$Br$_2$]$^+$ is olive green to maroon (the colors appear to be different under natural and artificial lighting), while *trans*-[Co(en)$_2$Br$_2$]$^+$ is lime green. In the solid state, *cis* dibromo bromide is charcoal grey; finer crystals take on an olive-green color. The cis complex has been resolved with both α-bromo-D-(+)-cam-phor-*trans*-π-sulfonate and (+)[Co(en)(C$_2$O$_4$)$_2$]$^-$,[4] and the active salts are distinctly greener than those of the racemate. The solution absorption spectra of all the cis salts are identical; they contain none of the green trans isomer despite the dramatic color variation which is presumably associated with particle size and crystal structure. The visible absorption spectra in water (extrapolated to $t = 0$) showed $\epsilon_{554}^{max} = 110$ (cis) and $\epsilon_{658}^{max} = 51.8$ (trans). The spectra of the trans and particularly the cis isomer change quickly with time; the spectrum in N,N-di-methylformamide is a more reliable criterion of purity for the cis isomer (ϵ_{554}^{max} 119.0).

The cis ($t_{1/2}$ 12 minutes, 25°) and trans ($t_{1/2}$ 83 minutes) dibromo complexes hydrolyze rapidly in dilute acidic aqueous solution to give a mixture of *cis-* and *trans*-Co(en)$_2$(OH$_2$)Br]$^{2+}$.[4,5] These aquation reactions are about 5-10 times faster than those of the corresponding dichloro isomers.

D. *cis*-AQUABROMOBIS(1,2-ETHANEDIAMINE)COBALT(III) DIBROMIDE MONOHYDRATE

$$\textit{trans-}[Co(en)_2Br_2]Br + 2H_2O \longrightarrow \textit{cis-}[Co(en)_2(OH_2)Br]Br_2 \cdot H_2O$$

Procedure

trans-Dibromobis(1,2-ethanediamine)cobalt(III) bromide (15.0 g) is dissolved in hot water (50 mL, 85°), and the temperature is maintained for 10 minutes. The original deep-green solution becomes violet grey and is immediately cooled to 10° using an ice/salt bath. Sodium bromide (30 g) is added with stirring. Violet crystals quickly appear together with a little unchanged green *trans*-[Co(en)$_2$Br$_2$]Br. After a further 15 minutes at 5° while stirring, the product is collected and washed with ethanol (5°, 3 × 30 mL) followed by diethyl ether (3 × 30 mL). The crude *cis*-[Co(en)$_2$(OH$_2$)Br]Br$_2$·H$_2$O (9.8 g; 63% yield) is freed of the trans dibromo impurity by a single recrystallization from water (80 mL, 20°). The filtered solution is cooled to 5°, and cold hydrobromic acid (40 mL) is added. After 30 minutes on ice, the violet crystals are collected and washed and dried as above. Yield 6.8 g, 44%.

Anal. Calcd. for *cis*-[Co(C$_4$H$_{16}$N$_4$)(OH$_2$)Br]Br$_2$·H$_2$O: C, 10.56; H, 4.43; N, 12.32; Br, 52.71. Found: C, 10.81; H, 4.54; N, 12.25; Br, 52.99.

E. *trans*-AQUABROMOBIS(1,2-ETHANEDIAMINE)COBALT(III) DITHIONATE MONOHYDRATE

$$trans\text{-}[Co(en)_2Br_2]\,Br + HgO + 2H_2SO_4 \longrightarrow trans\text{-} +$$
$$cis\text{-}[Co(en)_2(OH_2)Br]^{2+} + 2HSO_4^- + HgBr_2\downarrow$$
$$trans\text{-} + cis\text{-}[Co(en)_2(OH_2)Br]^{2+} + S_2O_6^{2-} + H_2O \longrightarrow$$
$$trans\text{-}[Co(en)_2(OH_2)Br]\,S_2O_6\cdot H_2O\downarrow$$

Procedure

trans-Dibromobis(1,2-ethanediamine)cobalt(III) bromide (8.4 g, 20 mmole) is added to a mixture of water (15 mL) and ice (5 g), and the suspension is kept at 5° in an ice/salt bath while well stirred. A previously prepared solution of mercuric oxide (red form, AnalaR, 4.56 g, 22 mmole) in a slight excess of aqueous sulfuric acid (45 mmole; 2.5 mL of 18 M in 15 mL water) and which is chilled to 5° is added *dropwise* over 5 minutes to the cobalt complex suspension while it is vigorously stirred. Initially, some of the trans dibromo complex precipitates as the hydrogen sulfate salt, together with mercuric bromide, but as the reaction proceeds, it redissolves and ultimately a deep maroon-burgundy solution results. After a total of 15 minutes, the solution is filtered to remove $HgBr_2$ which is washed with water (2 × 5 mL). To the filtrate and washings, maintained at 5° and stirred, is added freshly precipitated sodium dithionate dihydrate (7.3 g, 30 mmole).* Green-grey crystals of *trans*-$[Co(en)_2(OH_2)Br]$-$S_2O_6\cdot H_2O$ appear shortly. Stirring at 5° is continued for 10 minutes to complete the dissolution of $Na_2S_2O_6\cdot 2H_2O$ and the crystallization of the product. The complex is filtered, washed with methanol/ice (1:1, 2 × 10 mL), methanol (2 × 20 mL), and finally diethyl ether. The yield of air-dried product is 4.3 g (47% based on total cobalt, or approximately 60% based on the amount of available trans aquabromo ion in the cis/trans mixture). This complex is nearly pure *trans*-$[Co(en)_2(OH_2)Br]\,S_2O_6\cdot H_2O$, perhaps contaminated with a little $Na_2S_2O_6\cdot 2H_2O$. A single recrystallization by filtering a fresh aqueous solution (300 mL, 20°) into an ice/salt bath and quickly adding cold ethanol or methanol (1.5 L) affords fluffy pale green-grey needles (2.8 g; 31% overall yield). These are collected after 30 minutes at 5° and washed and dried as above. Some trans isomer is lost through isomerization to the cis form which remains in solution.

Anal. Calcd. for *trans*-$[Co(C_4H_{16}N_4)(OH_2)Br]\,S_2O_6\cdot H_2O$: Co, 12.95; C, 10.55;

---

*$Na_2S_2O_6\cdot 2H_2O$, as ordinarily purchased, consists of large crystals which are slow to dissolve. The rapid addition of acetone (1 L) to a hot filtered aqueous solution (100 g in 300 ml water at 85°) produces fine, white needles which are more suitable. These are collected, washed with acetone and diethyl ether, and dried in air.

H, 4.43; N, 12.31; S, 14.09. Found: Co, 12.94; C, 10.77; H, 4.55; N, 12.25; S, 14.21 %.

Properties

The bromide salt of the cis aquabromo ion is much less soluble than the corresponding trans salt. For the dithionate salts the converse is true. The cis complex is violet and the trans is grey green. As is usual for cis and trans isomers, the visible absorption spectra are quite distinct. The first ligand field band for the trans ion is weaker, and it appears at lower energies: $\epsilon_{531}^{max} = 91.0$, $\epsilon_{444}^{min} = 19.0$ (cis); $\epsilon_{606}^{max} = 35.5$, $\epsilon_{515}^{min} = 10.4$ (trans). The absorption at 530 nm provides a sensitive measure of isomeric purity ($\epsilon_{530}^{cis} = 91.0$, $\epsilon_{530}^{trans} = 11.4$). The spectral data given are extrapolated to zero time.

In dilute acid solution at $25°$, the cis and trans ions readily interconvert ($t_{1/2}$ 55 minutes), and the cis form predominates at equilibrium (74%). The aquation of the coordinated bromide ion is much slower than the isomerization reaction.[5] The cis–(+) isomer has been resolved as (+)-[Co(en)$_2$(CO$_3$)]$^+$, and it racemizes ($t_{1/2}$ 4.1 hours, $25°$, 0.1 M HClO$_4$) by conversion to the optically inactive trans isomer.[4]

F. *trans*-AQUACHLOROBIS(1,2-ETHANEDIAMINE)COBALT(III) DITHIONATE MONOHYDRATE

$$\text{trans-}[Co(en)_2Cl_2]\,Cl + HgO + 2H_2SO_4 \longrightarrow \text{trans-} +$$

$$\text{cis-}[Co(en)_2(OH_2)Cl]^{2+} + HgCl^+ + Cl^- + 2HSO_4^-$$

$$\text{trans-} + \text{cis-}[Co(en)_2(OH_2)Cl]^{2+} + S_2O_6^{2-} + H_2O \longrightarrow$$

$$\text{trans-}[Co(en)_2(OH_2)Cl]\,S_2O_6 \cdot H_2O\downarrow$$

Procedure

trans-Dichlorobis(1,2-ethanediamine)cobalt(III) chloride is prepared from [Co(en)$_2$(CO$_3$)]Cl (150 g) and concentrated hydrochloric acid (400 mL), as described for the dibromo analog in Section 28-B. Oven drying (100°, 2 hours) yields 143 g (92%). The material is recrystallized from a filtered, saturated aqueous solution by adding an equal volume of methanol and then diluting carefully with a 10-fold volume excess of acetone. After 1 hour at 5°, the fine grass-green needles are collected, washed with two portions each of acetone and then diethyl ether, and air dried. Recovery is nearly quantitative.

trans-Dichlorobis(1,2-ethanediamine)cobalt(III) chloride (11.4 g, 40mmole) is dissolved in water (25 mL), ice (10 g) is added, and the mixture is chilled to 5°. A solution of HgO (9.5 g, 45mmole) in water (50 mL) and sulfuric acid (18 M, 5.3 mL, 95mmole) is cooled to 5° and added dropwise to the green complex solution over 5 minutes. The initially precipitated *trans*-[Co(en)$_2$Cl$_2$]HSO$_4$ dissolves to a gray-black solution which is filtered. The addition of recrystallized Na$_2$S$_2$O$_6$·2H$_2$O (14.6 g, 60mmole) and stirring for a further 15 minutes yields a grey-green crystalline precipitate. This is collected, washed (methanol/ice, 1:1, 2 × 10 mL; methanol, 2 × 10 mL; diethyl ether, 2 × 10 mL), and dried in air. Yield 7.9 g, 48% based on total cobalt, or about 70% based on the available trans aquachloro isomer. The material is recrystallized as described for the bromo analog in Section 28-E to give green-grey fluffy needles. Recovery 4.6 g, 58%; overall yield 28%.

Anal. Calcd. for *trans*-[Co(C$_4$H$_{16}$N$_4$)(OH$_2$)Cl]S$_2$O$_6$·H$_2$O: Co, 14.25; C, 11.70; H, 4.91; N, 13.65; Cl, 8.63; S, 15.61. Found: Co, 14.31; C, 11.85; H, 4.88; N, 13.48; Cl, 8.68; S, 15.88.

Properties

These are very similar to those of the bromo analogs described in Section 28-E, except that the isomerization and racemization rates are slower.[4] The visible absorption spectrum of the trans isomer in dilute acid (extrapolated to $t = 0$) showed: $\epsilon_{587}^{max} = 31.1$, $\epsilon_{510}^{min} = 9.6$, $\epsilon_{439}^{max} = 30.9$. The tail of the very strong charge-transfer absorption associated with the Co—Br moiety obscures this latter band of the bromo analog. The absorption at 510 nm where $\epsilon_{510}^{cis} = 85.5$ is a sensitive measure of the isomeric purity.

Acknowledgments

Support for this work from the Australian Research Grants Committee is gratefully acknowledged.

References

1. J. Springbørg and C. E. Schäffer, *Inorg. Synth.*, **14**, 63 (1973).
2. A. Werner, *Ann. Chem.*, **386**, 1 (1912).
3. G. B. Kauffman, *Coordination Chem. Rev.*, **15**, 1 (1975).
4. W. G. Jackson and A. M. Sargeson, *Inorg. Chem.*, **17**, 1348 (1978).
5. C. G. Barraclough, R. W. Boschen, W. W. Fee, W. G. Jackson, and P. T. McTigue, *Inorg. Chem.*, **10**, 1994 (1971).
6. P. J. Staples and M. L. Tobe, *J. Chem. Soc.*, 4812 (1960).
7. R. D. Hargens, W. Min, and R. C. Henney, *Inorg. Synth.*, **14** 79 (1973).
8. F. P. Dwyer, A. M. Sargeson, and I. K. Reid, *J. Am. Chem. Soc.*, **85**, 1215 (1963).
9. J. W. Vaughn and R. D. Lindholm, *Inorg. Synth.*, **9**, 163 (1967).

29. TRIS(2,2'-BIPYRIDINE)RUTHENIUM(II) DICHLORIDE HEXAHYDRATE

Submitted by JOHN A. BROOMHEAD* and CHARLES G. YOUNG*
Checked by PAM HOOD†

The tris(2,2'-bipyridine)ruthenium(II) complex cation along with its substituted derivatives is an important species in the study of electron-transfer and likely solar energy conversion reactions. It was first prepared in low yield by pyrolysis of ruthenium trichloride with the ligand.[1] An improved yield was obtained by reflux of these reagents in ethanol,[2] but the method suffers from the long reaction time (72 hours). Another synthesis[3] uses $K_4[Ru_2Cl_{10}O] \cdot H_2O$ which itself requires a separate preparation. The method described here uses commercial ruthenium trichloride, is of high yield, and takes about 1 hour. It may also be used to prepare analogous 1,10-phenanthroline and related diimine ligand complexes.

The Nature of Commercial Hydrated Ruthenium Trichloride

Samples of $RuCl_3 \cdot xH_2O$ may contain Ru(IV), various oxo- and hydroxychloro-complexes, and nitrosyl species. Also, considerable variation in reaction times and product distribution may accompany the use of samples from different sources.[4] In this synthesis, consistently high yields of $[Ru(bipy)_3]Cl_2$ can be obtained if the $RuCl_3 \cdot xH_2O$ is oven treated prior to use.

Drying Procedure

Commercial $RuCl_3 \cdot xH_2O$ is dried in an oven at 120° for 3 hours. It is then finely ground in a mortar and returned to the oven for a further 1 hour prior to use. It is convenient to store the "dried" $RuCl_3$ at this temperature.

Procedure

$$RuCl_3 + 3C_{10}H_8N_2 \xrightarrow{NaH_2PO_2/H_2O} [Ru(C_{10}H_8N_2)_3]Cl_2 \cdot 6H_2O$$

"Dried" $RuCl_3$ (0.4 g, 1.93 mmole), 2,2'-bipyridine (0.9 g, 5.76 mmole), and

---

*Department of Chemistry, Faculty of Science, Australian National University, Canberra, A.C.T. 2600, Australia.
†Department of Chemistry, Oklahoma State University, Stillwater, OK 74048.

water (40 mL) are placed in a 100-mL flask fitted with a reflux condenser. Freshly prepared sodium phosphinate (sodium hypophosphite) solution (2 mL) is added and the mixture heated at the boil for 30 minutes. [The sodium phosphinate solution is prepared by the careful addition of sodium hydroxide pellets to about 2 mL 31% phosphinic acid (hypophosphorous acid) until a slight cloudy precipitate is obtained. Phosphinic acid is then added dropwise until the precipitate just redissolves.]

During reflux, the initial green solution changes to brown and finally orange. It is filtered to remove traces of undissolved material and potassium chloride (12.6 g) added to the filtrate to precipitate the crude product. The solution and solid are then heated at the boil to give a deep-red solution which on cooling to room temperature yields beautiful, red platelike crystals. These are filtered off, washed with ice-cold 10% aqueous acetone (2 × 5 mL) and acetone (30 mL), and air dried. The yield is 1.15 g (80%). The product may be recrystallized from boiling water ($\sim$ 2.8 mL/g) and then air dried.

Anal. Calcd. for $C_{30}Cl_2H_{36}N_6O_6Ru$. C, 48.13; Cl, 9.47; H, 4.85; N, 11.22. Found: C, 48.30; Cl, 9.62; H, 4.85; N, 11.27.

Properties

Aqueous solutions of $[Ru(bipy)_3]Cl_2 \cdot 6H_2O$ have two characteristic absorption maxima at 428 nm (shoulder ϵ = 11,700) and 454 nm (ϵ = 14,000) which have been assigned to metal ligand charge-transfer transitions.[5] The charge-transfer excited state is relatively long-lived in solution (lifetime $\sim$600 nsec),[6] and the luminescence spectrum (λ_{max} 600 nm) has been assigned to heavy-metal perturbed triplet-singlet phosphorescence of the excited state.[7,8] The complex has been resolved into its optical enantiomers using the iodide antimonyl(+)tartrate salt. Also the redox potentials of various derivatives have been measured.[9]

References

1. F. H. Burstall, *J. Chem. Soc.*, 173 (1936).
2. R. A. Palmer and T. S. Piper, *Inorg. Chem.*, **5**, 864 (1966).
3. F. P. Dwyer, *J. Proc. Roy. Soc. N.S.W.*, **83**, 134 (1949).
4. J. M. Fletcher, W. E. Gardner, E. W. Hooper, K. R. Hyde, F. M. Moore, and J. L. Woodhead, *Nature*, **199**, 1089 (1963).
5. J. P. Paris and W. W. Brandt, *J. Am. Chem. Soc.*, **81**, 5001 (1959).
6. C.-T. Lin, W. Böttcher, M. Chou, C. Creutz, and N. Sutin, *J. Am. Chem. Soc.*, **98**, 6536 (1976).
7. J. N. Demas and G. A. Crosby, *J. Mol. Spectrosc.*, **26**, 72 (1968).
8. F. E. Lytle and D. M. Hercules, *J. Am. Chem. Soc.*, **91**, 253 (1969).
9. W. W. Brandt, F. P. Dwyer, and E. C. Gyarfas, *Chem. Rev.*, **54**, 959 (1954).

30. ALKYLATED POLYAMINE COMPLEXES OF PALLADIUM(II)

Submitted by F. S. WALKER,* S. N. BHATTACHARYA,† and C. V. SENOFF*
Checked by KENT BAREFIELD‡ and GARY M. FREEMAN‡

The isomeric alkylated polyamines N,N-bis[2-(dimethylamino)ethyl]N',N'-dimethyl-1,2-ethanediamine(tris[2-(dimethylamino)ethyl]amine), $N(CH_2CH_2NMe_2)_3$ (normally abbreviated as trenMe$_6$) and N,N'-bis[2-(dimethylamino)ethyl] -N,N'-dimethyl-1,2-ethanediamine, $Me_2NC_2H_4NMeC_2H_4NMeC_2H_4$-$NMe_2$ (normally abbreviated as trienMe$_6$) can be conveniently coordinated to palladium(II) to form [PdCl(trenMe$_6$)]Cl and [Pd(trienMe$_6$)]Cl$_2$, respectively, via an initial reaction between the amine and palladium(II) chloride at low pH followed by further reaction at high pH in the presence of Na_2CO_3. The covalently bonded chloride in [PdCl(trenMe$_6$)]Cl may be readily replaced by bromide, iodide, or thiocyanate ions by metathesis at room temperature with the appropriate sodium salt in chloroform. The general procedure described herein is based on that originally reported for the synthesis of chloro[N-[2-(dimethylamino)ethyl] -N,N',N'-trimethyl-1,2-ethanediamine]palladium(II) hexafluorophosphate [PdCl(dienMe$_5$)](PF)$_6$,[1] and can be used to prepare other palladium(II) complexes containing an alkylated polyamine.[2]

A. [N,N-BIS[2-(DIMETHYLAMINO)ETHYL]-N',N'-DIMETHYL-1,2-ETHANEDIAMINE]CHLOROPALLADIUM(II) CHLORIDE, [PdCl(trenMe$_6$)]Cl

$$PdCl_2 + 2HCl \xrightarrow{\Delta} H_2PdCl_4$$

$$H_2PdCl_4 + N(CH_2CH_2NMe_2)_3 \xrightarrow{\Delta}$$

$$\{PdCl[N(CH_2CH_2NMe_2)_2(CH_2CH_2NMe_2H)]\}Cl_2 + HCl$$

$$\{PdCl[N(CH_2CH_2NMe_2)_2(CH_2CH_2NMe_2H)]\}Cl_2 + Na_2CO_3 \longrightarrow$$

$$[PdCl[N(CH_2CH_2NMe_2)_3]]Cl + CO_2 + NaOH + NaCl$$

*Department of Chemistry, University of Guelph, Guelph, Ontario, Canada, N1G 2W1.
†Department of Chemistry, Lucknow University, Lucknow, India.
‡Department of Chemistry, Georgia Institute of Technology, Atlanta, GA 30332.

Procedure

Palladium(II) chloride (1.0 g, 5.6 mmole) is added to a solution of 30 mL water and 1.0 mL concentrated HCl contained in a 100-mL, round-bottomed flask. The resulting mixture is heated under reflux until all the palladium(II) chloride dissolves. The dark reddish-brown solution is then allowed to cool to room temperature, and trenMe$_6$[3] (1.65 g, 7.1 mmole) is slowly added. The resulting yellow solution is then heated at reflux for 30 minutes and allowed to cool again to room temperature. Small portions of Na$_2$CO$_3$ are carefully added to the solution until the pH is adjusted to approximately 11.0 as measured by using a "colorpHast" indicator stick [the checkers report that careful addition of 25% NaOH solution works equally well.]. The solvent is removed by means of a rotoevaporator and the yellow residue is dried in vacuo at room temperature. The palladium complex is extracted from the residue by the successive use of four 25-mL portions of hot chloroform. The combined chloroform extracts are filtered to remove insoluble materials and are evaporated to near dryness on the rotoevaporator. Acetone (30 mL) is added to precipitate the product, which is collected by filtration and dried in vacuo at room temperature. The yield is 1.2 g (52%) [the checkers report a yield of 2.15 g (93%) when ether is used to precipitate the product which is somewhat soluble in acetone.].

Anal. Calcd. for $C_{12}H_{30}Cl_2N_4Pd$: C, 35.35; H, 7.42; N, 13.74; Cl, 17.39. Found: C, 34.95; H, 7.45; N, 13.37; Cl, 17.24.

Properties

N,N-bis[2-(dimethylamino)ethyl]-*N',N'*-dimethyl-1,2-ethanediamine] chloropalladium(II) chloride is a yellow, diamagnetic solid that melts with decomposition at 210°. It is somewhat hygroscopic and is best stored in a moisture-free atmosphere. It is very soluble in water, chloroform, and dichloromethane but insoluble in most other common laboratory solvents.

The ^{1}H nmr spectrum in D$_2$O exhibits a single resonance at 2.52 ppm for the methyl protons with DSS, the sodium salt of 2,2-dimethyl-2-silapentane-5-sulfonic acid as reference. The electronic spectrum for an aqueous solution is characterized by an absorption maximum at 358 nm (ϵ = 7.4 Lmole^{-1} cm^{-1}) in H$_2$O. Other physical properties are described in the literature [the checkers report $\delta(CH_3)$ = 2.6 ppm with λ_{max} = 336 nm, ϵ = 650 Lmol^{-1} cm^{-1}].

B. [*N,N*-BIS[2-(DIMETHYLAMINO)ETHYL]-*N',N'*-DIMETHYL-1,2-ETHANEDIAMINE]IODOPALLADIUM(II) IODIDE, [PdI(trenMe$_6$)]I

$$[PdCl(trenMe_6)]Cl + 2NaI \longrightarrow [PdI(trenMe_6)]I + 2NaCl$$

Procedure

A 0.5 g (1.2 mmole) sample of [PdCl(trenMe$_6$)]Cl is dissolved in chloroform (15 mL) in a 100-mL, round-bottomed flask and to this solution freshly dried (24 hours in vacuo at 60-70°) NaI (0.45 g, 3.0 mmole) is added. The dead-space above the solution is purged with dry N$_2$, the flask is stoppered, and the mixture is stirred at room temperature for 24 hours during which time it changes from yellow to violet-red in color. The solution is then filtered by gravity and the residue washed with three 5-mL portions of chloroform. The filtrate and washings are combined and the solvent removed with the aid of a rotoevaporator. The resulting residue is dissolved in a minimum amount of acetone (< 5 mL), and 60 mL hexane is added with constant stirring to precipitate [PdI(trenMe$_6$)]I as a finely divided, dark-yellow solid. The complex is collected by filtration, washed with hexane, and dried in vacuo at room temperature. The yield is 0.64 g (88%). [the checkers obtained an orange solution, not violet which may be a result of oxidation of I$^-$, on addition of NaI. Triturination with hexane prevented oil formation]

Anal. Calcd. for C$_{12}$H$_{30}$I$_2$N$_4$Pd: C, 24.41; H, 5.12; N, 9.48; I, 42.99. Found: C, 24.41; H, 5.05; N, 9.60; I, 42.89.

Properties

[*N,N*-bis[2-(dimethylamino)ethyl]-*N',N'*-dimethyl-1,2-ethanediamine]iodopalladium(II) iodide is a yellow, diamagnetic solid that melts with decomposition at 160°. It is not very hygroscopic but should be stored in a moisture-free atmosphere. It is soluble in water, chloroform, acetone, and dichloromethane but insoluble in most other common laboratory solvents. The ^{1}H nmr spectrum in D$_2$O exhibits a single resonance at 2.68 ppm for the methyl protons (DSS reference). The electronic spectrum for an aqueous solution is characterized by an absorption maximum at 414 nm (ϵ = 625 L mole^{-1} cm^{-1}). Other physical properties are reported in the literature.[5]

C. **[*N,N*-BIS[2-(DIMETHYLAMINO)ETHYL]-*N',N'*-DIMETHYL-1,2-ETHANEDIAMINE]BROMOPALLADIUM(II) BROMIDE, [PdBr(trenMe$_6$)]Br**

$$[PdCl(trenMe_6)]Cl + 2NaBr \longrightarrow [PdBr(trenMe_6)]Br + 2NaCl$$

Procedure

The procedure given in Section 30-B is followed, except for the replacement of NaI by NaBr (0.31 g, 3.0 mmole), and the reaction time is increased to 48 hours. After the reaction mixture is filtered, it is concentrated on the rotoevaporator and hexane is added to precipitate the complex. The yield is 0.43 g (72%).

Anal. Calcd. for $C_{12}H_{30}Br_2N_4Pd$: C, 29.03; H, 6.09; N, 11.28; Br, 32.19. Found: C, 28.99; H, 6.15; N, 11.12; Br, 32.22.

Properties

[*N*,*N*-bis[2-(dimethylamino)ethyl] -*N'*,*N'*-dimethyl-1,2-ethanediamine] bromo-palladium(II) bromide is a yellow diamagnetic solid. It melts at 210° with decomposition. It is very hygroscopic and must be handled and stored in a moisture-free atmosphere. It is soluble in water, chloroform, and dichloromethane but insoluble in most other common laboratory solvents. The ^{1}H nmr spectrum in D_2O exhibits a single resonance at 2.60 ppm for the methyl protons. The electronic spectrum obtained on an aqueous solution is characterized by an absorption maximum at 352 nm (ϵ = 792 L mole^{-1} cm^{-1}).

D.   [*N*,*N*-BIS[2-(DIMETHYLAMINO)ETHYL]-*N'*,*N'*-DIMETHYL-1,2-ETHANEDIAMINE](THIOCYANATO-*N*)PALLADIUM(II) THIOCYANATE, [Pd(NCS)(trenMe$_6$)]SCN

$$[PdCl(trenMe_6)]\,Cl + 2NaSCN \longrightarrow [Pd(NCS)(trenMe_6)]\,SCN + 2NaCl$$

Procedure

A sample of [PdCl(trenMe$_6$)]Cl (1.0 g, 2.4 mmole) is dissolved in chloroform (20 mL) in a 100-mL, round-bottomed flask, and to this solution is added freshly dried (24 hours in vacuo at 60-70°) NaSCN (0.50 g, 6.0 mmole). The dead-space above the solution is purged with dry N_2, the flask is stoppered, and the mixture is stirred at room temperature for 48 hours, during which time a red oil forms. [the checkers observed no red oil.] The supernatant is decanted and the oily, red residue is washed with several 5-mL portions of chloroform. The supernatant and washings are combined and the solvent removed with the aid of a rotoevaporator. The residue is then thoroughly dried in vacuo at room temperature and then dissolved in a minimum amount of dry acetone. Yellow crystals form after 6-8 hours at room temperature which are collected by filtration and dried in vacuo in the presence of P_4O_{10}. The yield is variable due to problems encountered during the extraction of the oily complex from the NaCl formed, but is in the range 35-55%.

Anal. Calcd. for $C_{14}H_{30}N_6S_2Pd$: C, 37.14; H, 6.68; N, 18.55; S, 14.16. Found: C, 36.82; H, 6.55; N, 18.56; S, 14.25.

Properties

[*N,N*-bis[2-(dimethylamino)ethyl] -*N′,N′*-dimethyl-1,2-ethanediamine] (thiocyanato-*N*)palladium(II) thiocyanate is a diamagnetic, yellow solid. It is very hygroscopic and should be stored in a moisture-free atmosphere. The palladium atom is four-coordinate in the solid state.[5] The ^{1}H nmr spectrum in D_2O exhibits 2 singlet resonances at 2.56 and 2.63 ppm for the methyl protons (DSS as reference). The electronic spectrum in H_2O is characterized by 2 absorption maxima at 263 nm (ϵ = 7200 L mole^{-1} cm^{-1}) and 325 nm $(\epsilon$ = 1800 L mole^{-1} cm^{-1}). The infrared spectrum (Nujol) exhibits ν(CN) bands at 2105 and 2060 cm^{-1}.

E. [*N,N′*-BIS[2-(DIMETHYLAMINO)ETHYL] -*N,N′*-DIMETHYL-1,2-ETHANE-DIAMINE]PALLADIUM(II) BIS(HEXAFLUOROPHOSPHATE), [Pd(trienMe$_6$)](PF$_6$)$_2$

$$PdCl_2 + 2HCl \xrightarrow{\Delta} H_2PdCl_4$$

$$H_2PdCl_4 + Me_2N(C_2H_4NMe)_2C_2H_4NMe_2 \xrightarrow{\Delta}$$
$$[PdCl(Me_2N(C_2H_4NMe)_2C_2H_4NMe_2H)] Cl_2 + HCl$$

$$[PdCl(Me_2N(C_2H_4NMe)_2C_2H_4NMe_2H)] Cl_2 + Na_2CO_3 \longrightarrow$$
$$[Pd(Me_2N(C_2H_4NMe)_2C_2H_4NMe_2)] Cl_2 + CO_2 + NaOH + NaCl$$

$$[Pd(Me_2N(C_2H_4NMe)_2C_2H_4NMe_2)] Cl_2 + 2NH_4PF_6 \longrightarrow$$
$$[Pd(Me_2N(C_2H_4NMe)_2C_2H_4NMe_2)] (PF_6)_2 + 2NH_4Cl$$

Procedure

Palladium(II) chloride (0.50 g, 2.8 mmole) is added to a solution of 20 mL water and 1.0 mL concentrated HCl contained in a 100-mL, round-bottomed flask. The resulting mixture is heated at reflux until all the palladium(II) chloride dissolves. The dark reddish-brown solution is then allowed to cool to room temperature, and *N,N′*-bis[2-(dimethylamino)ethyl] -*N′,N′*-dimethyl-1,2-ethanediamine (1.0 mL), is slowly added to the solution. (Ames Laboratories, Inc., 200 Rock Lane, Milford, CT 06460) Small portions of Na_2CO_3 are then added to the orange solution until the pH is adjusted to approximately 11.0 as measured by using a "colorpHast" indicator stick. The solvent is then removed by means of a rotoevaporator, leaving a yellow oily residue.

The residue is added to 100 mL absolute ethanol and undissolved solids are removed by filtration. The filtrate is evaporated to a volume of about 20 mL, and to this is added a hot aqueous solution of ammonium hexafluorophosphate (1.0 g in 10.0 mL water) to precipitate the product, which is collected by filtration. Yield is 0.45 g (50%). The crude product can be recrystallized from a minimum volume of hot acetone.*

Anal. Calcd. for $C_{12}H_{30}F_{12}N_4P_2Pd$: C, 22.99; H, 4.79; N, 8.94. Found: C, 22.85; H, 5.05; N, 8.88.

Properties

[N,N'-Bis[2-(dimethylamine)ethyl]-N',N'-dimethyl-1,2-ethanediamine] palladium(II) bis(hexafluorophosphate) is an air-stable, yellow solid which is soluble in acetonitrile and insoluble in water and other common laboratory solvents. It melts at 235°. The ^1H nmr spectrum in CD_3CN is characterized by a large number of sharp lines with a more intense resonance at 2.7 ppm (TMS reference). The electronic spectrum obtained on an aqueous solution shows a single absorption maximum at 350 nm ($\epsilon = 594$ L mole^{-1} cm^{-1}). Other properties are reported in the literature.[1]

References

1. W. H. Baddley and F. Basolo, *J. Am. Chem. Soc.*, 88, 2944 (1966).
2. S. N. Bhattacharya, C. V. Senoff, and F. Walker, *Syn. React. Inorg. Metal. Org. Chem.*, 9, 5 (1979).
3. M. Ciampolini and N. Nardi, *Inorg. Chem.*, 5, 41 (1966).
4. C. V. Senoff, *Inorg. Chem.*, 17, 2320 (1978).
5. G. Ferguson and M. Parvez, *Acta Crystallogr.*, submitted for publication.

*The checkers report that the following procedure gives a much improved yield. Palladium(II) chloride (0.5 g, 2.8 mmole) is added to a solution of 10 mL water and 0.5 mL concentrated HCl. The mixture is heated on a steam bath until all of the palladium chloride dissolves (~10 min). The solution is cooled to room temperature and N,N'-bis[2-(dimethylamino)ethyl]-N',N'-dimethyl-1,2-ethanediamine (1 g, 4.3 mmole) is slowly added to the solution. The pH of the solution is slowly brought to about 11 with addition of 25% NaOH solution and then filtered to remove any insoluble material. The resulting clear yellow solution is heated to 50-60 °C and 1.2 g (7.3 mmole) ammonium hexafluorophosphate dissolved in 5-6 mL water is slowly added with vigorous stirring. The mixture is cooled in an ice bath to complete the precipitation. The yellow product is collected by filtration and dried in vacuo (0.01 torr). Yield is 1.5 g (86%). The product may be recrystallized by dissolving it in 20 mL of a 1:1 v/v acetone-water mixture containing 0.3 g ammonium hexafluorophosphate. After gravity filtration the solution is reduced in volume to about 10 mL on a steam bath. Crystallization of the product is completed by cooling the mixture in an ice bath. The product is collected by filtration, washed with 2 mL of cold water, and dried as described above. Recovery is essentially quantitative. Use of the small quantity of HCl is important. Some Cl⁻ contamination may otherwise result.

31. TETRAHYDROFURAN COMPLEXES OF SELECTED EARLY TRANSITION METALS

Submitted by L. E. MANZER[*]
Checked by JOE DEATON,[†] PAUL SHARP,[†] and R. R. SCHROCK[†]

Few reagents are available for the preparation of organometallic and coordination complexes of the early transition metals. The anhydrous metal halides often lead to disproportionation reactions, and the nitrile derivatives,[1] $MX_n(NCR)_y$, are not suitable because of the reactivity of the nitrile with many other reagents. The procedures described below provide simple, high-yield routes to the tetrahydrofuran (THF) complexes of selected, early transition metal halides.[2] They have been useful for the synthesis of a wide variety of organometallic complexes.[3]

The anhydrous metal halides were obtained from Alfa-Ventron Corp., Danvers, MA 01923. (Only the very best metal halides should be used to produce the yields reported.) All solvents were dried over 4 Å molecular sieves and sparged with nitrogen prior to use.

■ NOTE: *All reactions and manipulations should be performed under an atmosphere of dry nitrogen either in a dry box or using Schlenk-tube techniques.*

A. TETRACHLOROBIS(TETRAHYDROFURAN)TITANIUM(IV), TiCl₄(thf)₂

$$TiCl_4 + 2 C_4H_8O \longrightarrow TiCl_4(OC_4H_8)_2$$

Procedure

A 125-mL Erlenmeyer flask equipped with a magnetic stirring bar is charged with 5.0 g (26mmole) titanium tetrachloride dissolved in 50 mL dichloromethane. Anhydrous tetrahydrofuran (7.62 g, 0.10 mole) is added dropwise, and the solution is stirred at room temperature, under nitrogen, for 15 minutes. (■ Caution. *Do not attempt to add anhydrous TiCl₄ directly to tetrahydrofuran as a violent, exothermic reaction occurs.*) Dry pentane (50 mL) is added, and the solution is chilled to −25° for 1 hour. A bright-yellow solid is collected on a medium fritted funnel and washed with 25 mL dry pentane. The pentane wash plus another 50 mL dry pentane are added to the mother liquor, which is

*Central Research and Development Dept., E. I. du Pont de Nemours and Co., Experimental Station, Wilmington, DE 19898. Current address: E. I. du Pont, Victoria, TX 77901.
†Department of Chemistry, Massachusetts Institute of Technology, Cambridge, MA 02139.

then chilled to $-25°$ for 1 hour. A second crop of yellow solids is collected. The total yield is 8.10 g (91.8%).

Anal. Calcd. for $C_8H_{16}Cl_4O_2Ti$: C, 28.75; H, 4.79; Cl, 42.46; Ti, 14.34. Found: C, 28.98; H, 4.83; Cl, 42.09; Ti, 14.53.

Properties

The product is a bright-yellow, crystalline solid which is air sensitive. It has a melting point of 126-128°. Its infrared spectrum contains a strong, sharp peak at 990 cm^{-1} and a strong broad peak at 825-845 cm^{-1}.

B.   TETRACHLOROBIS(TETRAHYDROFURAN)ZIRCONIUM(IV), ZrCl$_4$(thf)$_2$

$$ZrCl_4 + 2C_4H_8O \longrightarrow ZrCl_4(OC_4H_8)_2$$

■ NOTE: *See the recommendation above concerning the use of an inert atmosphere.*

A 500-mL Erlenmeyer flask equipped with a magnetic stirring bar is charged with a suspension of 23.3 g (100mmole) zirconium tetrachloride in 300 mL dichloromethane at room temperature. (The checkers note that insolubles are formed if the ZrCl$_4$ is not pure, producing a lower apparent yield.) Anhydrous tetrahydrofuran (14.42 g, 200mmole) is added dropwise, causing an exothermic reaction strong enough to reflux the dichloromethane. (■ **Caution.** *Do not attempt to add anhydrous ZrCl$_4$ directly to tetrahydrofuran as a violent, exothermic reaction occurs with significant darkening of the solid.*) As the tetrahydrofuran is added, the zirconium tetrachloride dissolves giving a slightly turbid, colorless solution at the end of the addition. The solution is filtered through a medium fritted funnel into a 1-L flask. To the filtrate is added 250 mL dry pentane, and this solution is chilled to $-25°$ for 2 hours. The product is collected on a medium fritted funnel, washed with 50 mL dry pentane, and dried in vacuo. The product is a white, crystalline solid weighing 34.20 g (90.7%).

Anal. Calcd. for $C_8H_{16}Cl_4O_2Zr$: C, 25.45; H, 4.24; Cl, 37.59; Zr, 24.18 Found: C, 26.08; H, 4.51; Cl, 38.20; Zr, 23.68.

Properties

The product is a white, crystalline solid which is air sensitive. It melts with decomposition at 170-171°. Its infrared spectrum contains a sharp, strong peak at 990 cm^{-1} and a strong, broad peak at 820-840 cm^{-1}.

C. TETRACHLOROBIS(TETRAHYDROFURAN)HAFNIUM(IV), HfCl₄(thf)₂

$$HfCl_4 + 2OC_4H_8 \longrightarrow HfCl_4(OC_4H_8)_2$$

A 125-mL Erlenmeyer flask equipped with a magnetic stirring bar is charged with a suspension of 5.0 g (15.6mmole) hafnium tetrachloride in 50 mL dichloromethane. Anhydrous tetrahydrofuran (4.50 g, 15.6mmole) is added dropwise, and the hafnium tetrachloride dissolves as the tetrahydrofuran is added. (■ **Caution.** *Do not attempt to add anhydrous HfCl₄ directly to tetrahydrofuran as a violent, exothermic reaction occurs with significant darkening of the solid.*) After stirring for 15 minutes, a white solid precipitates. This solid is collected on a medium fritted funnel and washed with 25 mL dry pentane. A total of 50 mL dry pentane is added to the mother liquor, which is then chilled to −35° for 1 hour. A second crop of white solid precipitates; this is collected and washed with 50 mL dry pentane. The total yield is 6.35 g (87.6%). (The checkers' yield was 50%; they recommend immediate filtration to remove insolubles.)

Anal. Calcd. for $C_8H_{16}Cl_4O_2Hf$: C, 20.67; H, 3.44; Cl, 30.52; Hf, 38.42. Found: C, 21.22; H, 3.49; Cl, 31.43; Hf, 39.50.

Properties

The product is a white, crystalline solid which is air sensitive. It melts at 187°, and the infrared spectrum contains a sharp peak of medium intensity at 990 cm^{-1} and a strong, broad peak at 820 cm^{-1}.

D. TRICHLOROTRIS(TETRAHYDROFURAN)TITANIUM(III), TiCl₃(thf)₃

$$TiCl_3 + 3C_4H_8O \longrightarrow TiCl_3(OC_4H_8)_3$$

Procedure

A 500-mL, single-neck, round-bottom flask equipped with a reflux condenser and magnetic stirring bar is charged with 200 mL anhydrous tetrahydrofuran and 10.0 g (65mmole) titanium trichloride. The solution is heated to reflux for 22 hours, then allowed to cool to room temperature. The resulting pale-blue crystals are collected on a medium fritted funnel and washed with 50 mL dry pentane. The pentane wash plus another 50 mL dry pentane are added to the mother liquor to give a second crop of pale-blue solids. The total yield is 21.96 g (91.4%).

Anal. Calcd. for $C_{12}H_{24}Cl_3O_3Ti$: C, 38.89; H, 6.48; Cl, 28.68; Ti, 12.92. Found: C, 38.69; H, 6.61; Cl, 29.31; Ti, 12.81.

Properties

The product is a pale-blue, air-sensitive, crystalline solid. It slowly changes color but does not melt below 250°. Its infrared spectrum contains a strong, sharp peak at 1010 cm^{-1} and a strong, broad peak at 850 cm^{-1}.

E. TRICHLOROTRIS(TETRAHYDROFURAN)VANADIUM(III), VCl$_3$(thf)$_3$

$$VCl_3 + 3\ OC_4H_8 \longrightarrow VCl_3(OC_4H_8)_3$$

Procedure

A 500-mL, single-neck, round-bottom flask equipped with a reflux condenser and magnetic stirring bar is charged with 200 mL anhydrous tetrahydrofuran and 10.0 g (63 mmole) vanadium trichloride. The solution is refluxed for 22 hours, then chilled to −80° with a Dry ice bath and filtered through a medium fritted funnel. The pink crystals are washed, under nitrogen, with 50 mL dry pentane and dried in vacuo under reduced pressure to give 20.8 g (87.6%).

Anal. Calcd. for $C_{12}H_{24}Cl_3O_3V$: C, 38.57; H, 6.42; Cl, 28.46; O, 12.85; V, 13.63. Found: C, 38.72; H, 6.43; Cl, 29.89; O, 13.30; V, 14.55.

Properties

The product is a pink, crystalline solid which is air sensitive. It slowly changes color but does not melt below 250°. Its infrared spectrum contains a strong, sharp peak at 1010 cm^{-1} and a broad, strong peak at 850 cm^{-1}.

F. TETRACHLOROBIS(TETRAHYDROFURAN)NIOBIUM(IV), NbCl$_4$(thf)$_2$

$$NbCl_4 + Al \xrightarrow{CH_3CN} NbCl_4(NCCH_3)_3 + AlCl_3$$

$$NbCl_4(NCCH_3)_3 + OC_4H_8 \longrightarrow NbCl_4(OC_4H_8)_2 + 3\ NCCH_3$$

A 125-mL Erlenmeyer flask equipped with a magnetic stirring bar is charged with 1.6 g aluminum powder and 50 mL dry acetonitrile. To this suspension is added with stirring 4.8 g (18 mmole) niobium tetrachloride. After stirring under nitrogen for 4 hours, the solution is filtered through a medium fritted funnel giving a pale, yellow-green powder and a dark, orange-brown liquid. The aceto-nitrile mother liquor is removed by rotary evaporation to give a reddish-brown solid. This solid is suspended in 50 mL anhydrous tetrahydrofuran and stirred at room temperature for 16 hours. The solution is then filtered, giving a yellow powder which is washed with 25 mL anhydrous tetrahydrofuran and 50 mL dry pentane, and dried in vacuo under reduced pressure. The total yield is 4.71 g, or 69.1%.

Anal. Calcd. for $C_8H_{16}Cl_4O_2Nb$: C, 25.36; H, 4.25; Cl, 37.43; O, 8.44. Found: C, 25.15; H, 3.78; Cl, 36.86; O, 8.44.

Properties

The product is a pale-yellow, air-sensitive powder which melts at 145°C after slow color change. Its infrared spectrum contains a sharp peak of medium intensity at 990 cm^{-1} and a strong, broad peak at 820 cm^{-1}.

G. TRICHLOROTRIS(TETRAHYDROFURAN)SCANDIUM(III), ScCl₃(thf)₃

$$ScCl_3 \cdot 6H_2O \xrightarrow{\text{thf/SOCl}_2} ScCl_3(OC_4H_8)_3 + SO_2 + HCl$$

Procedure

A 250-mL, three-neck, round-bottom flask equipped with a reflux condenser, a magnetic stirring bar, and an addition funnel, is charged with 10.0 g ScCl₃·6H₂O suspended in 100 mL anhydrous tetrahydrofuran. By way of the addition funnel, 50 mL sulfinyl chloride is added dropwise. This causes an exothermic reaction, bringing the tetrahydrofuran to reflux. After the addition, the solution is held at reflux for 6 hours. The tetrahydrofuran and excess sulfinyl chloride are then removed, using a rotary evaporator, leaving behind white crystals. These crystals are then washed with 200 mL diethyl ether and vacuum dried, giving 13.69 g, or 98% yield. *Anal.* Calcd.: C, 39.21; H, 6.57; Sc, 12.23; Cl, 28.93. Found: C, 39.23; H, 6.60; Sc, 12.56; Cl, 29.53.

Properties

The product is a white, crystalline solid which does not melt below 250°. An infrared spectrum of the solid contains a strong, sharp peak at 1010 cm^{-1} and a stronger but broader peak at 855 cm^{-1}.

References and Notes

1. *Inorg. Synth.*, **12**, 225 (1970).
2. See J. R. Dilworth and R. L. Richards, Inorg. Synth, **20**, 121 (1980), for similar syntheses of $MoCl_4(thf)_2$ and $MoCl_3(thf)_3$.
3. J. Chatt and A. G. Wedd, *J. Organometal. Chem.*, **27**, C15 (1971).
4. L. E. Manzer, *Inorg. Chem.*, **16**, 525 (1977).
5. L. E. Manzer, *J. Am. Chem. Soc.*, **99**, 276 (1977).

Chapter Five

SOLID STATE

32. ONE-DIMENSIONAL TETRACYANOPLATINATE COMPLEXES CONTAINING PLATINUM-ATOM CHAINS

Submitted by JACK M. WILLIAMS*

Metal-chain complexes containing stacked square-planar tetracyanoplatinate groups, $[Pt(CN)_4]^{2-}$, are currently of high interest because of their one-dimensional (very anisotropic!) metallic properties. Complexes of this type contain metal-atom chains and often possess a characteristic brilliant, metallic luster. They may be synthesized by oxidation using chemical or electrolytic techniques.[1] Although these compounds often appear metallic, they may also be semiconductors. These complexes differ in their Pt-Pt intrachain separations, degree of partial oxidation of the platinum atom ($Pt^{2.1-2.4}$), electrical conductivity, and metallic color.[2] Compounds in this series which contain platinum atoms in a nonintegral oxidation state are known as "partially oxidized tetracyanoplatinate" (POTCP) complexes. Some complexes also possess a metallic luster but are not metallic, as is the case for $Tl_4(CO_3)[Pt(CN)_4]$ (see below).

■ **Caution.** *Because of the extremely poisonous nature of cyanide, chlorine and thallium, and the very corrosive nature of HF, these steps should be carried out in a well-ventilated fume hood using protective gloves and clothing and face shield. Should any solution containing HF spill on gloves or clothing, they should be removed immediately. At no time should any solutions containing HF*

*Chemistry Division, Argonne National Laboratory, Argonne, IL 60439.

141

be exposed to glass. The possible presence of hydrazoic acid during the acidification of CsN_3 requires that a well-ventilated hood be used and that protective gloves, face shield, and barrier be used.

Materials

The $Cs_2[Pt(CN)_4] \cdot H_2O$, $Rb_2[Pt(CN)_4] \cdot 1.5 H_2O$, $[C(NH_2)_3]_2[Pt(CN)_4] \cdot xH_2O$, and $K_2[Pt(CN)_4] \cdot 3H_2O$ starting materials for the following preparations are synthesized as described by Maffly et al.,[3] Koch et al.,[4] Cornish and Williams,[5] and Abys et al.,[6] respectively. A general synthetic route for the improved synthesis of tetracyanoplatinate starting materials using $Ba[Pt(CN)_4] \cdot 4H_2O$ is described by Maffly and Williams.[7] The CsCl used in procedure A is 99+% pure (Matheson, Coleman, and Bell and Alfa Division, Ventron Corporation). The RbCl used in procedure B is 99.9% pure. In procedure D, the anhydrous KF is Alpha Products 99%, the HF is Baker reagent grade, and the H_2O_2 is Fischer Scientific (30%). In procedure E, the CsN_3 is Eastman Kodak No. 8556 (99.9%). In procedure F, the Tl_2SO_4 and Tl_2CO_3 are Alfa Division, Ventron Corporation, 99.5% and ultrapure, respectively. All other chemicals are ACS reagent grade. Throughout the procedures, distilled water is used, and all reactions involving the use of HF are carried out in polyethylene beakers.

A. CESIUM TETRACYANOPLATINATE CHLORIDE (2:1:0.30), $Cs_2[Pt(CN)_4]Cl_{0.3}$

Submitted by ANTHONY F. PRAGOVICH, JR.,* CHRISTOPHER C. COFFEY,* LEAH E. TRUITT,* and JACK M. WILLIAMS†
Checked by STEVEN T. MASUO and JOEL S. MILLER‡

Chemical Method

$$Cs_2[Pt^{II}(CN)_4] \cdot H_2O + Cl_2 \xrightarrow{H_2O} Cs_2[Pt^{IV}(CN)_4Cl_2] + H_2O$$

$$3Cs_2[Pt^{IV}(CN)_4Cl_2] + 17 Cs_2[Pt^{II}(CN)_4] \cdot H_2O \xrightarrow{CsCl}$$

$$20 Cs_2[Pt(CN)_4]Cl_{0.3} + 17 H_2O$$

*Research participants sponsored by the Argonne Division of Educational Programs: Anthony F. Pragovich, Jr. from Illinois State University, Normal, IL; Christopher C. Coffey from Mercyhurst College, Erie, PA; and Leah E. Truitt from Washington College, Chestertown, MD.
†Chemistry Division, Argonne National Laboratory, Argonne, IL 60439.
‡Occidental Research Corporation, Irvine, CA 92713.

Electrolytic Method

$$Cs_2[Pt(CN)_4] \cdot H_2O + CsCl + HCl \xrightarrow{\text{1.00 V dc}} Cs_2[Pt(CN)_4]Cl_{0.3} + H_2O$$

Procedure—Chemical Method.

1. Initially 1 mL of a saturated solution of $Cs_2[Pt(CN)_4] \cdot H_2O^3$ is added to 4 mL water. Chlorine gas is bubbled through this solution for a period of 15 minutes. During this period, the color of the solution changes from clear and colorless to a clear yellow. The mixture is then heated in a boiling water bath for 3 minutes to drive off excess chlorine.

This Pt(IV)-containing solution is then allowed to evaporate in a desiccator until crystallization occurs. Chemical analysis of the crystals indicate that they are $Cs_2[Pt(CN)_4Cl_2]$.

Anal. Calcd. for $Cs_2[Pt(CN)_4Cl_2]$: C, 7.56; N, 8.81; Cl, 11.15; H, 0.0; O, 0.0. Found:[8] C, 7.5; N, 8.82; Cl, 10.77; H, 0.0; O, 0.73.

Normally, the Pt(IV) salt is not isolated when carrying out the next procedure. Instead, the solution is used directly in the next step.

2. Cesium chloride (CsCl) [3.89 g (23.1mmole)] is dissolved in 10 mL water. This CsCl solution is added to the Pt(IV) solution from step 1 along with 10 mL water to insure dissolution of the CsCl.

The mixture is transferred into a 100-mL polyethylene beaker, and 5 mL of a saturated solution of $Cs_2[Pt(CN)_4] \cdot H_2O$ is added. Upon addition, a black suspension may form. Mild heating in a boiling water bath, after adding an additional 30 mL water, causes dissolution.

The solution is allowed to evaporate in a desiccator maintained at $25°$. Within 48 hours, metallic bronze-colored crystals form. In two different preparations the checkers of this synthesis obtained a 72 and 77% yield of the final product in this step.

Anal. Calcd. for $Cs_2[Pt(CN)_4]Cl_{0.30}$: C, 8.35; N, 9.73; Cl, 1.85; H, 0.0; O, 0.0. Found[8]: C, 8.34; N, 9.49; Cl, 1.89; H, 0.23; O, 1.89.

Emission spectrographic analysis indicates the product to be of high purity and confirmed the presence of Pt and Cs only.[8] The following is a list of impurities[8]: 0.02% Al, 0.004% Ba, a trace of Ca (<0.01%), a faint trace of Mg/Cu, Si(<0.001%), 0.07% K, and 0.1% Na.

Procedure—Electrolytic Method

Initially, 2.1 mL of a saturated solution of $Cs_2[Pt(CN)_4] \cdot H_2O^3$ is added to 1.04 mL of a saturated solution of CsCl.[9] Upon addition, a white precipitate forms. The volume of the solution is increased to 10 mL with stirring, and dissolution is complete. The solution is then acidified with 0.1 mL concentrated HCl.

This solution is transferred to a 50-mL electrolytic cell. The electrolytic cell is prepared by drilling two small holes on opposite sides of a 50-mL polyethylene beaker ~5 mm from the bottom of the beaker and cementing two platinum wires in the holes using epoxy cement. The platinum wires should be approximately parallel and separated by 3-5 mm.

The cell is connected to a variable-voltage power supply.[10] A voltage of 1.00 V dc is applied. Initially, fuzzy, bronze-colored, needlelike crystals form; longer crystals form in a 2-4 week period. The crystals are collected by suction filtration through a medium-porosity fritted glass filter. The crystals are washed with four 1-mL portions of ice-cold ($0°$) water and allowed to dry in the air on filter paper. The checkers of this synthesis obtained a 62% yield via the electrolytic method. Larger crystals can be prepared by using one-half the concentration of Pt as stated in the electrolytic method. In this case, no white precipitate forms, however, a smaller yield is obtained.

Anal. Calcd. for $Cs_2[Pt(CN)_4]Cl_{0.30(2)}$: C, 8.35; N, 9.73; Cl, 1.85; H, 0.0. Found[8]: C, 8.39, 8.43, 8.33, 8.29, avg. 8.36; N, 9.83, 9.43, 9.73, 9.21, avg. 9.55; Cl, 1.64, 1.75, 1.59, 1.76, 1.68, 1.75, 1.79, 1.99, avg. 1.74; H, 0.22, 0.34, 0.30, 0.27, avg. 0.28.

Emission spectrographic analysis indicated the product to be of high purity and confirmed the presence of only the metals Pt and Cs.[8]

Physical Properties

The compound $Cs_2[Pt(CN)_4]Cl_{0.3}$ prepared by either method above forms fine, needlelike crystals which are metallic bronze in color. X-Ray diffraction data on both crystals yield identical cell constants and indicate the unit cell to be body-centered tetragonal[11]: a = 13.176(2) Å, c = 5.718(1) Å, and V = 992.7 Å.^{3} Experimental confirmation of the calculated density of 3.889 g/cm^3 was not possible for lack of a suitable solvent.

Both thermogravimetric and elemental analysis indicate the possibility of a trace (less than 0.25 H_2O per Pt atom) of water, which may account for the presence of hydrogen in the observed analytical results.[12] As noted by the checkers, the best fit of the analytical data yields a formula of $Cs_2[Pt(CN)_4]Cl_{0.29} \cdot 0.36H_2O$. However, single-crystal X-ray structure analysis[11] does not show the presence of water within the crystal lattice unless the partially occupied Cl sites are occasionally replaced by H_2O.

The first 12 reflections of the powder pattern correspond to the following *d*-spacings (Å)[8]: 9.28(vs), 6.47(m), 4.62(m), 4.12(vs), 3.06(m-w), 2.92(s), 2.83(s), 2.76(s) 2.49(w), 2.57(m-w), 2.42(m), 2.33(vs).

B. RUBIDIUM TETRACYANOPLATINATE CHLORIDE (2:1:0.30) HYDRATE, $Rb_2[Pt(CN)_4]Cl_{0.30} \cdot 3H_2O$

Submitted by CHRISTOPHER C. COFFEY* and JACK M. WILLIAMS†

$$Rb_2[Pt(CN)_4] \cdot 1.5H_2O + Cl_2 \xrightarrow[RbCl]{H_2O} Rb_2[Pt(CN)_4]Cl_{0.30} \cdot 3H_2O$$

Procedure‡

Initially, 0.5 g (1 mmole) $Rb_2[Pt(CN)_4] \cdot 1.5H_2O^4$ is added to 15 mL H_2O with heating (approx. 70°) and stirring. Approximately 1 mL liquefied chlorine gas, cooled in an acetone/dry ice bath, is added cautiously. (■ **Caution.** *Chlorine gas is a hazardous material. Handle with care!*) The mixture is then stirred for 20 minutes to drive off the excess chlorine, and then an additional 2.5 g (5 mmole) $Rb_2[Pt(CN)_4] \cdot 1.5H_2O$ is added. The solution is allowed to stir for another 15 minutes and then evaporated by boiling to 10 mL and finally cooled in an ice bath. The resulting crystals are isolated. An X-ray diffraction powder pattern demonstrates that the crystals are monoclinic $Rb_{1.75}[Pt(CN)_4] \cdot 1.5H_2O$.

To obtain the desired product, the crystals are then redissolved in 25 mL water, and 5.98 g (50 mmole) RbCl is added to make the solution $2\,M$ in RbCl. This solution is then placed in a desiccator and allowed to evaporate for ~24 hours until the desired crystals appear. The $Rb_2[Pt(CN)_4]Cl_{0.3} \cdot 3H_2O$ may also be prepared by adding 5 mL concentrated HCl to 6 mL of a saturated solution of $Rb_2[Pt(CN)_4] \cdot xH_2O$ which is then electrolyzed (1.0 V dc).

The yield was 1.0311 g (2.1 mmole), or 34.3% based on the original $Rb_2[Pt(CN)_4] \cdot 1.5H_2O$. The final product reported here can only be produced when it is crystallized from a solution containing excess ($>2\,M$) RbCl. Both chemical[8] and neutron diffraction crystal structure analyses[13] confirm the composition of this complex as reported here.

*Research participant sponsored by the Argonne Division of Educational Programs and from Mercyhurst College, Erie, PA.
†Chemistry Division, Argonne National Laboratory, Argonne, IL 60439.
‡Editor's Note: This procedure has not been checked specifically (for *Inorg. Synth.*) but is similar to other procedures submitted by Dr. Williams which have been checked.

Physical Properties

Crystals of $Rb_2[Pt(CN)_4]Cl_{0.3}\cdot 3H_2O$ have a metallic copper luster and appear to be relatively stable toward hydration and dehydration. They may be stored indefinitely in a sealed glass container. Neutron diffraction[13] analysis indicates that $Rb_2[Pt(CN)_4]Cl_{0.3}\cdot 3H_2O$ crystallizes in the tetragonal system with unit cell dimensions $a = 10.142(6)$ Å and $c = 5.801(4)$ Å and the observed density is $2.98(1)$ g/cm^3.

C. GUANIDINIUM TETRACYANOPLATINATE (HYDROGEN DIFLUORIDE)(2:1:0.27) HYDRATE, $[C(NH_2)_3]_2[Pt(CN)_4]$ $(FHF)_{0.27}\cdot 1.8H_2O$

Submitted by CHRISTOPHER C. COFFEY,* CURTISS L. WHITE, JR.,* and JACK M. WILLIAMS†

$$100[C(NH_2)_3]_2[Pt(CN)_4]\cdot xH_2O + 54\ HF \xrightarrow[1.5\ V\ (D.C.)]{H_2O}$$

$$100[C(NH_2)_3]_2[Pt(CN)_4](FHF)_{0.27}\cdot 1.8H_2O + 14\ H_2$$

Procedure‡

Initially, 1.60 g $[C(NH_2)_3]_2[Pt(CN)_4]\cdot xH_2O$[5] is dissolved in 5 mL concentrated HF (28.9 M) in a 50-mL polyethylene beaker. The solution is transferred to an electrolysis cell which is constructed from a polyethylene beaker with two holes drilled ~2 cm beneath the top of the solution. Two platinum wires ~25 mm long are cemented in the two holes with epoxy cement and separated by ~5 mm to form the electrodes. A constant potential is applied to the electrodes using a 1.5-V dc source.[10] Crystal growth is observed anytime between 1 and 5 days. One week after initial crystal formation, the crystals are suction filtered through a plastic funnel using Whatman qualitative filter paper. The material is then washed with two 5-mL portions of cold water and allowed to air dry. The yield is 1.46 g (80%) based on $[C(NH_2)_3]_2[Pt(CN)_4]\cdot xH_2O$ being anhydrous.

Anal. Calcd. for $[C(NH_2)_3]_2[Pt(CN)_4][HF_2]_{0.27}\cdot 1.8H_2O$: Pt, 43.33; C, 16.01; N, 31.12; H, 3.53; F, 2.34. Found[8]: C, 15.71, 15.94, 15.94; N, 31.33, 31.46, 30.91; H, 2.93; F, 2.35, 2.17, 2.31, 2.15, 2.44, 2.46; Pt, 46.15.

*Research participants sponsored by the Argonne Division of Educational Programs: Christopher C. Coffey from Mercyhurst College, Erie, PA, and Curtis L. White, Jr., from Brown University, Providence, RI.
†Chemistry Division, Argonne National Laboratory, Argonne, IL 60439.

Thermogravimetric analyses indicate that the compound contains 1.8 mole water. The tentatively accepted platinum oxidation state is +2.27 based on the fluorine content of the compound and has an estimated standard deviation of ±0.02. Iodine thiosulfate titrations indicate a platinum oxidation state of +2.26.[8]

Properties

$[C(NH_2)_3]_2[Pt(CN)_4][HF_2]_{0.27} \cdot 1.8H_2O$ forms shiny, black crystals which belong to the body-centered tetragonal crystal glass. From preliminary single-crystal X-ray diffraction data, the unit cell lattice constants are $a = b = 15.5(1)$ Å and $c = 5.8(1)$ Å. The first 10 reflections of the X-ray diffraction powder pattern correspond to the following d-spacings (Å)[8]: 10.96(vs), 7.85(vs), 5.48(s), 4.90(s), 4.24(vw), 3.63(vw), 3.44(m), 3.13(vw), 3.03(w), and 2.89(w). The compound appears to be relatively stable toward hydration and dehydration and may be stored over a saturated solution of NaCl. However, the compound should be stored in a plastic container and refrigerated to avoid the loss of HF.

D. POTASSIUM TETRACYANOPLATINATE (HYDROGEN DIFLUORIDE) (2:1:0.30) HYDRATE, $K_2[Pt(CN)_4](FHF)_{0.30} \cdot 3H_2O$

Chemical Method

See basic chemical equations given in Section 32-A.1 and substitute K for Cs, KF for CsCl, and H_2O_2 for Cl_2.

Electrolytic Method

See basic chemical equation in Section 32-A.2 and use V = 1.5 (dc) and substitute K for Cs and KF for CsCl.

Procedure[‡]—Chemical Method

Initially, 1.55 g (3.59 mmole) of $K_2[Pt(CN)_4] \cdot 3H_2O$[7] is dissolved in 6 mL of water followed by 1.51g (26.0mmole) of anhydrous KF and then 3 mL concentrated HF(aq), all in a 50-mL polyethylene beaker. Three drops (40 drops/mL) (0.7mmole) 30% H_2O_2 are then added. The solution is swirled and allowed to evaporate at room temperature until the beaker is filled with crystals, being cautious not to allow drying to occur. The crystals are collected in a

[‡]Editor's Note: This procedure has not been checked specifically (for *Inorg. Synth.*) but is similar to other procedures submitted by Dr. Williams which have been checked.

plastic funnel, washed twice with 6 mL cold H_2O and allowed to air dry for 10 minutes. Based on the original $K_2[Pt(CN)_4] \cdot 3H_2O$, the 0.5-g yield represents 32%.

Anal. Calcd. for $K_2[Pt(CN)_4](FHF)_{0.3} \cdot 3H_2O$: K, 17.94; Pt, 44.75; C, 11.02; N, 12.85; H, 1.27; F, 2.61; O, 9.54. Found[8]: K, 17.35; Pt, 44.55; F, 2.87. Found[8]: C, 10.98, 10.87, 11.30; N, 12.51, 12.69, 12.62; H, 1.05, 1.07, 1.04; F, 2.51, 2.64, 2.75, 2.80, 2.67; O, 9.91; Halogen other than F, 0.0.

Procedure—Electrolytic Method

As in the chemical method, 1.55 g (3.59 mmole) $K_2[Pt(CN)_4] \cdot 3H_2O$ is dissolved in 6 mL water. KF (anhydrous) (1.51 g, 26.0mmole) is added and the KF dissolves leaving a fine, white suspension. Finally, 3 mL concentrated HF(aq) (24 *M*) is added to acidify the solution and promote solution of the white suspension. The clear solution is then transferred to the electrolytic cell, which consists of a 50-mL polyethylene beaker with two small opposing holes drilled ∼6 mm from the bottom. Two platinum electrodes ∼25 mm in length are held in place using epoxy cement and with the electrode tips separated by ∼2 mm.

After ∼2 hours of electrolysis[10] at 1.5 V, the product, having filled the bottom of the beaker, is filtered in a plastic funnel. The crystals are then washed with two ∼6-mL portions of cold water and allowed to air dry. The yield of 1.4-1.5 g (3.2-3.4mmole) represents 88-94% based on original $K_2[Pt(CN)_4] \cdot 3H_2O$. Using X-ray diffraction powder patterns, it is established that the products of both syntheses given above are identical.[8] The platinum oxidation state of 2.29(1) is established by iodine-thiosulfate titrations.[8]

Properties

The compound $K_2[Pt(CN)_4](FHF)_{0.3} \cdot 3H_2O$ forms tetragonal crystals which have a reddish-bronze metallic luster. The full X-ray diffraction crystal structure analysis has been reported.[14] The powder pattern indicates that $K_2[Pt(CN)_4](FHF)_{0.3} \cdot 3H_2O$ is isostructural with $K_2[Pt(CN)_4]Cl_{0.3} \cdot 3H_2O$, which also has a platinum oxidation state of +2.30. The first 10 reflections of the powder pattern correspond to the following d-spacings (Å)[8]: 9.671 (vs), 8.770 (M), 8.104 (M), 6.884 (s), 4.878 (s), 4.651 (w), 4.362 (vs), 4.264 (vs), 4.076 (vw), and 3.944 (vw).

$K_2[Pt(CN)_4](FHF)_{0.3} \cdot 3H_2O$ appears to be relatively stable toward hydration and dehydration and may be stored safely over a saturated solution of NaCl which has a relative humidity of 75.1% at 25°. However, it must be stored in a plastic container to avoid possible reactions with glass.

E. CESIUM TETRACYANOPLATINATE AZIDE (2:1:0.25) HYDRATE, $Cs_2[Pt(CN)_4](N_3)_{0.25} \cdot xH_2O$

Submitted by DAVID A. VIDUSEK* and JACK M. WILLIAMS†
Checked by STEVEN T. MASUO‡ and JOEL S. MILLER‡

$$Cs_2[Pt(CN)_4] \cdot H_2O + 0.25\ CsN_3 + 0.25\ H_2O \xrightarrow[\text{1.0 V dc}]{H_2SO_4}$$

$$Cs_2[Pt(CN)_4](N_3)_{0.25} \cdot xH_2O + 0.25\ CsOH + 0.125\ H_2O$$

Procedure

A saturated solution of $Cs_2[Pt(CN)_4] \cdot H_2O^3$ is prepared by adding 1.26 g (2.16 mmole) $Cs_2[Pt(CN)_4] \cdot H_2O$ to 9 mL distilled water. It may be necessary to warm the solution slightly in order to dissolve the $Cs_2[Pt(CN)_4] \cdot H_2O$. The solution is then acidified to a pH of approximately 5.5 with the addition of 0.2 mL of 9 M sulfuric acid. It should be noted that the checkers of this synthesis find that crystals grow faster at lower pH and that the addition of 0.2 mL of 9 M H_2SO_4 resulted in an initial pH of 2.1. The sulfuric acid is added in small increments to minimize the possible generation of HCN gas. Upon addition of the sulfuric acid, a white suspension is formed which does not dissipate. To this mixture, 1.5 g (8.57 mmole) of CsN_3 is added. Upon addition, the CsN_3 changes color from white to red and then dissolves rapidly, leaving the white suspension within the solution. The solution is then suction filtered through a fine-porosity sintered glass filter and the filtrate is saved. The clear solution is transferred to an electrolytic cell consisting of a 50-mL polyethylene beaker with two small opposing holes drilled 5 mm from the bottom. Two platinum electrodes 25 mm in length are set in place through the holes using epoxy cement and with the electrode tip separation being approximately 6 mm.

After 72 hours of electrolysis at 1.0 V dc, the reddish-bronze crystals are suction filtered through a medium-porosity sintered glass filter. The crystals are then washed with two 3-mL portions of cold water and allowed to air dry. A yield of 1.15 g (1.96 mmole) represents essentially a 91% yield based on the original $Cs_2[Pt(CN)_4] \cdot H_2O$.

*Research participant sponsored by the Argonne Division of Educational Programs from Illinois State University, Normal, IL 61761.
†Chemistry Division, Argonne National Laboratory, Argonne, IL 60439.
‡Occidental Research Corporation, Irvine, CA 92713.

TABLE I Chemical Analyses of $Cs_2[Pt(CN)_4](N_3)_{0.25} \cdot xH_2O$[a]

Element	Percentage Composition							
	Sample I	Sample 2	Sample 3	Sample 4	Sample 5	Sample 6	Average of Samples	Theoretical
C	8.34	8.23	8.23	8.22	8.16	8.59	8.28(1)	8.28
	8.17	8.52	8.14	8.29	8.18	8.33		
H	0.30	0.30	0.19	0.21	0.12	0.15		
	0.11	0.54	0.39	0.22				
O		1.71	1.63	1.58		1.20	0.25(41)	
							1.53(1)	
N	11.64	11.78	11.79	11.69	10.85	11.23	11.47(2)	11.47
	11.51	11.46	11.59	11.66	10.83	11.49		
	11.50	11.60	11.61	11.50	11.13	11.52		
Pt[g]				32.56			32.56	33.62
Cs[g]				44.54			44.54	45.81

[a]The authors are indebted to the checkers of this synthesis for a computer analysis of the data presented in Table I which indicates that if the azide:Pt ratio is not fixed, and the N_3^- and H_2O content are varied, the best fit to the data yields the formulation $Cs_2[Pt(CN)_4](N_3)_{0.23} \cdot 0.36H_2O$ for this complex.

Chemical analyses and theoretical values are reported in Table I.[8] The calculated platinum oxidation state, based on chemical analysis and crystallographic analysis, is +2.25. An X-ray diffuse scattering analysis also indicates that the Pt oxidation state is 2.25(2).[15]

Emission spectrographic analyses[8] indicate the product to be of high purity and containing impurities as follows: Al, 0.01%; Ba, 0.2%; Cu, faint trace; Li, 0.0001%; Rb, 0.005%; Mg, faint trace; and Na, 0.006%.

Attempts to prepare this compound using a 10-fold excess of 30% H_2O_2 rather than by electrolytic oxidation have failed to produce a similar product.

Physical Properties

The compound $Cs_2[Pt(CN)_4](N_3)_{0.25} \cdot xH_2O$ forms long, needle-shaped crystals which have a distinctive reddish-bronze metallic luster. The complex crystallizes in the tetragonal space group $P\bar{4}b2$ with cell constants as determined from single crystal X-ray diffractometer measurements of: a = 13.089(2) Å, b = 13.089(2) Å, c = 5.754(1) Å, and V = 985.78 Å.[3] The calculated density is 3.96 g/cm^3 based on four formula weights per unit cell. The first 10 reflections of the powder pattern correspond to the following d-spacings (Å)[8]: 9.2126(9), 6.6074(1), 4.5753(6), 4.0953(10), 2.9122(2), 2.8666(3), 2.7143(4), 2.3441(8), 2.2325(7), and 2.1809(5). The platinum-platinum intrachain spacing in the compound is 2.88 Å. The $Cs_2[Pt(CN)_4](N_3)_{0.25} \cdot xH_2O$ exhibits a room temperature single crystal conductivity (four probe) of ~6 ohms^{-1}cm^{-1}.

F. CESIUM [HYDROGEN BIS(SULFATE)] TETRACYANOPLATINATE (3:0.46:1), $Cs_3[Pt(CN)_4](O_3SO \cdot H \cdot OSO_3)_{0.46}$

Submitted by DAVID A. VIDUSEK* and JACK M. WILLIAMS†
Checked by STEVEN T. MASUO‡ and JOEL S. MILLER‡

$$Cs_2[Pt(CN)_4] \cdot H_2O + CsN_3 + 0.8\ H_2SO_4 \xrightarrow[\text{0.6 V dc}]{H_2O}$$

$$Cs_3[Pt(CN)_4](O_3SO \cdot H \cdot OSO_3)_{0.46} + N_2 + NH_2OH + 0.2H^+$$

*Research participant sponsored by the Argonne Division of Educational Programs from Illinois State University, Normal, IL, 61761.
†Chemistry Division, Argonne National Laboratory, Argonne, IL, 60439.
‡Occidental Research Corporation, Irvine, CA, 92713.

Procedure

A cesium tetracyanoplatinate(II) monohydrate[3] 0.24 M solution is prepared by adding 1.26 g (2.15 mmole) $Cs_2[Pt(CN)_4] \cdot H_2O$ to 9 mL distilled water, and the mixture is warmed slightly (40-50°). Then, cesium azide (1.50 g, 8.57 mmole) is added to the colorless solution, which becomes reddish and then fades to white and rapidly dissolves. The solution is then acidified with 0.8 mL 9 M H_2SO_4 (pH $\cong$ 0.7). If the solution becomes turbid, it should be filtered through a fine-pore frit before using. Upon acidification, bubbles form and the sulfuric acid is added in small increments to minimize possible generation of HCN gas. The clear solution is transferred to an electrolytic cell consisting of a 50-mL polyethylene beaker with two small opposing holes drilled 5 mm from the bottom. Two platinum electrodes are set in place through the holes using epoxy cement and with an electrode tip separation of approximately 4 mm.

After 7 days of electrolysis at 0.6 V,[10] the brown, metallic crystals are suction filtered through a medium-porosity fritted filter, washed with two 3-mL portions of cold water, and allowed to air dry. A yield of 0.35 g (0.441 mmole) represents a 20% yield based on the original $Cs_2[Pt(CN)_4] \cdot H_2O$. The checkers of this procedure obtained a 15-16% yield in three different syntheses.

A preliminary crystallographic study indicates that the complex is isomorphous with $Rb_3[Pt(CN)_4](O_3SO \cdot H \cdot OSO_3)_{0.46} \cdot H_2O$.[16] The formulation of this complex as $Cs_3[Pt(CN)_4](O_3SO \cdot H \cdot OSO_3)_{0.46}$ as determined from the crystallographic analysis is in excellent agreement with the elemental analysis (Table II).

A platinum oxidation state of 2.27 is determined by iodine-thiosulfate titrations[8] assuming a MW of 786.78. The calculated platinum oxidation state, based on chemical analyses and crystallographic analysis, is 2.38, and we are unable to explain this discrepancy.

TABLE II   Chemical Analyses of $Cs_3[Pt(CN)_4](SO_4 \cdot H \cdot SO_4)_{0.46}$

Element	Sample 1		Sample 2		Mean and Std. Dev.	Theoretical
C	6.30	6.22	6.32	5.92	6.19(3)	6.10
N	6.96	7.22	7.04	7.00	7.06(1)	7.12
S	3.25	3.40	3.76	4.08	3.62(10)	3.74
O	11.22	10.88	9.28	9.55	10.23(69)	7.48
H	0.54	0.37	0.40	0.33	0.41(1)	0.46
Cs						50.71
Pt						24.78

Emission spectrographic analysis[8] indicates the product to be of high purity and containing impurities as follows: Rb, 0.02%; Ga, 0.005%; Ba, 0.005%; and a faint trace (0.01%) of Na and Al. Attempts at preparing this compound via a chemical preparation using a 3- or 10-fold excess of 30% H_2O_2 have failed to produce a similar product.

Physical Properties

The compound $Cs_3[Pt(CN)_4](O_3SO \cdot H \cdot OSO_3)_{0.46}$ forms triclinic crystals which have a brownish-bronze metallic luster. The cell constants as determined from single crystal X-ray diffractometer measurements are as follows: $a = 5.783(2)$ Å, $b = 9.571(8)$ Å, $c = 14.499(5)$ Å, $\alpha = 71.43(5)°$, $\beta = 81.93(3)°$, $\gamma = 73.19(5)°$, $V = 727.2(7)$ Å.[3] The calculated crystal density is 3.59 g/cm^3 based on 2 formula weights per unit cell.

The first 12 reflections of the X-ray diffraction powder pattern correspond to the following d-spacings (Å)[8]: 13.80, 8.76, 6.86. 6.39, 4.77, 4.57, 4.31, 4.14, 3.97, 3.81, 3.59, and 3.04.

The cesium[hydrogen bis(sulfate)] tetracyanoplatinate complex exhibits a room-temperature electrical conductivity (four probe technique) of approximately 113 ohms^{-1} cm^{-1}, while that of the Rb analog (Pt–Pt - 2.83 Å) is ~1500-2000 ohms^{-1} cm^{-1}.

G. TETRATHALLIUM CARBONATE TETRACYANOPLATINATE(II), $Tl_4(CO_3)[Pt(CN)_4]$

Submitted by LEAH E. TRUITT,* WADE A. FREEMAN,* and JACK M. WILLIAMS†
Checked by A. P. GINSBERG and C. R. SPRINKLE‡

Procedure

1. Dithallium Tetracyanoplatinate (II):

$$Ba[Pt(CN)_4] \cdot 4H_2O + Tl_2SO_4 \xrightarrow[70°]{H_2O} Tl_2[Pt(CN)_4] + BaSO_4 + H_2O$$

*Research participants sponsored by the Argonne Division of Education Programs: Ms. Leah E. Truitt from Washington College, Chestertown, MD, and Dr. Wade A. Freeman from the University of Illinois at Chicago Circle, Chicago, IL.
†Chemistry Division, Argonne National Laboratory, Argonne, IL 60439.
‡Bell Laboratories, 600 Mountain Avenue, Murray Hill, NJ 07974.

Initially, 7.0 g (13.8 mmole) $Ba[Pt(CN)_4] \cdot 4H_2O^3$ is dissolved with stirring in 25 mL hot (70-80°) water. To this is added a solution of 7.0 g (13.8 mmole) Tl_2SO_4 in 210 mL warm (50-60°) water, producing an immediate precipitation of $BaSO_4$. The temperature is maintained at 60° and stirring continued for 1 hour. The solution is then filtered through a fine-pore fritted glass filter. The $BaSO_4$ is washed with two 10-mL portions of hot (70-80°) water (see below). The filtrate is transferred to a boiling water bath and the volume is reduced to 200 mL. The solution is allowed to cool slowly for 30 minutes and then placed in an ice bath for 90 minutes to induce slow crystallization of $Tl_2[Pt(CN)_4]$. The white crystals are isolated on a medium-pore fritted glass filter and dried in the air on filter paper. Yield 7.1 g (~80% based on $Ba[Pt(CN)_4] \cdot 4H_2O$). The filtrate is returned to the boiling water bath and its volume is further reduced to 20 mL. It is again cooled, and a second crop of crystals is similarly isolated. Yield 0.2-0.4 g (1-3% based on $Ba[Pt(CN)_4] \cdot 4H_2O$). Total yield 7.0-7.49 g (75-80%). We are indebted to the checkers of this procedure who increased the yield in this step to 9.2 g (94%) by washing the $BaSO_4$ with two 50-mL portions of boiling water rather than the two 10-mL portions recommended by the authors.

Anal. Calcd. for $Tl_2[Pt(CN)_4]$: Tl, 57.74; Pt, 27.56; C, 6.79; N, 7.91; H, 0.0; O, 0.0. Found[8]: C, 6.91, 6.81; N, 8.05, 7.90; O, 0.65, 0.58; H, <0.01, <0.01. Found[8]: Tl, 57.88, 57.68; Pt, 27.78, 27.45; C, 6.43, 6.51; N, 7.59, 7.67; O, <0.5, <0.5; H, <0.05, <0.05. The presence of only the metals Tl and Pt is confirmed by emission spectrographic analysis.[8] Thermogravimetric analysis (25-100°) establishes that the compound is anhydrous.

Physical Properties

The compound $Tl_2[Pt(CN)_4]$ forms orthorhombic crystals which are pale yellow-white in color. The unit cell constants as determined from single-crystal X-ray diffraction are as follows: $a = 11.84$ Å, $b = 10.03$ Å, $c = 7.37$ Å, $V = 863.68$ Å.[3] The observed crystal density of 5.7 g/cm^3 agrees with the calculated density of 5.44 g/cm^3 based on 4 formula weights per unit cell. The first 12 reflections of the X-ray diffraction power pattern correspond to the following d-spacings (Å)[8]: 5.85 (w), 4.59 (vs), 3.88 (s), 3.65 (s), 3.39 (m), 3.10 (s-vs), 2.86 (m), 2.80 (w), 2.75 (w), 2.67 (w), 2.46 (s), and 2.32 (vs).

2.   Tetrathallium Carbonate Tetracyanoplatinate(II)[17]

$$Tl_2[Pt(CN)_4] + Tl_2CO_3 \xrightarrow{\text{H}_2\text{O}} Tl_4(CO_3)[Pt(CN)_4]$$

Initially 7.08 g (10mmole) $Tl_2[Pt(CN)_4]$ from reaction 1 is dissolved with stirring in 60 mL boiling water. A solution of 4.68 g (10mmole) of Tl_2CO_3 in 40 mL boiling water is added and stirring is continued until dissolution is complete.

After reducing the volume to 60 mL, or to the point at which a red precipitate begins to form, the reaction vessel is covered with a watch glass and then placed in a boiling water bath in a large, covered Dewar flask. The contents are allowed to cool gradually (about 2-3 days) to room temperature. Because of the extreme insolubility of the product in water below its boiling point, to obtain large crystals the boiling water bath must be of significant volume (3-4 L) and be between 98 and 100°, initially. This procedure yields large green crystals of the product, containing some chunks of white $Tl_2[Pt(CN)_4]$. Separation of the white $Tl_2[Pt(CN)_4]$ can be effected mechanically after collection by suction filtration on a fine-pore fritted glass filter. Alternatively, the $Tl_2[Pt(CN)_4]$ is removed by recrystallizing the mixture from a hot Tl_2CO_3 solution ($\sim5 \times 10^{-2}$ M) and cooling as just described. The large crystals so obtained (1-5 mm in length) are collected by suction filtration through a fine-pore fritted glass filter and dried in the air on filter paper. Yield 10.0 g ($\sim85\%$ based on $Tl_2[Pt(CN)_4]$).

Anal. Calcd. for $Tl_4(CO_3)[Pt(CN)_4]$: Tl, 69.48; Pt, 16.58; C, 5.10; N, 4.76; O, 4.08. Found[8]: Tl, 69.30, 69.46, 69.53; Pt, 16.68, 16.58; C, 4.97, 4.90; N, 4.40, 4.25; O, 4.17, 3.84. Found[8]: C, 5.13; N, 4.69; O, 4.49; H, <0.01.

Thermogravimetric analysis (25-100°) shows a <0.05% weight loss, indicating that $Tl_4(CO_3)[Pt(CN)_4]$ is anhydrous. Emission spectrographic analysis[8] indicates the product to be of high purity; it contains the metals Tl and Pt and impurities as follows; faint traces of Ca and Li (<0.001%). Iodine-thiosulfate titration studies[8] are negative, indicating *no* partial oxidation of Pt; therefore, Pt is present as $Pt^{2.0}$.

Physical Properties

The complex $Tl_4(CO_3)[Pt(CN)_4]$ forms tetragonal crystals which have a metallic gold-green luster in reflected light and a red color in transmitted light. The crystal structure has been determined.[18] The cell constants as determined from single-crystal X-ray diffractometer measurements are as follows: $a = 14.008(3)$ Å, $b = 14.008(3)$ Å, $c = 6.484(2)$ Å, and V = 1272.25 Å.^{3} The observed crystal density of 6.61(±1.1) g/cm^3 is comparable to the calculated density of 6.14 g/cm^3 based on 4 formula weights per unit cell. The first 10 reflections of the powder pattern correspond to the following d-spacings (Å)[8]: 4.37(s), 3.61(s), 3.19(m), 3.10(m), 2.77(s), 2.57(ms), 2.20(wm), 2.16(wm), 2.04(m), and

1.92(m). The $Tl_4(CO_3)[Pt(CN)_4]$ complex (avg. Pt—Pt = 3.242 Å) is a room-temperature semiconductor as expected from the Pt—Pt spacing,[2,18] even though it possesses a metallic luster.

Acknowledgments

This was performed under the auspices of the Office of Basic Energy Sciences, Division of Materials Sciences, of the Department of Energy. *By acceptance of this article, the publisher and-or recipient acknowledges the U.S. Government's right to retain a nonexclusive, royalty-free, license in and to any copyright covering this paper.*

References and Notes

1. For additional details and discussion, see J. M. Williams et al., *Inorg. Synth.*, **20**, 23 (1980).
2. J. M. Williams, A. J. Schultz, A. E. Underhill, and K. Carneiro, *Extended Linear Chain Compounds*, J. S. Miller (ed.), Plenum Press, New York, 1982, pp 73-118.
3. R. L. Maffly, J. A. Abys, and J. M. Williams, *Inorg. Synth.*, **19**, 6 (1979).
4. T. R. Koch, J. A. Abys, and J. M. Williams, *Inorg. Synth.*, **19**, 9 (1979).
5. T. F. Cornish and J. M. Williams, *Inorg. Synth.*, **19**, 10 (1979).
6. J. A. Abys, N. P. Enright, H. M. Gerdes, T. L. Hall, and J. M. Williams, *Inorg. Synth.*, **19**, 1 (1979).
7. R. L. Maffly and J. M. Williams, *Inorg. Synth.*, **19**, 112 (1979). Also see the correction by S. Kurtz, L. H. Truitt, R. C. Sundell, and J. M. Williams, *Inorg. Synth.*, **20**, 243 (1980).
8. Midwest Microlabs, Indianapolis, IN. This laboratory does not perform Pt or Cs analyses. When metal analyses are reported, they were performed by Galbraith Laboratory, Knoxville, TN. The presence of alkali metals, thallium, and platinum was confirmed by spectrographic analysis at Argonne National Laboratory. The authors wish to thank J. P. Faris and E. A. Huff for performing the spectrographic analyses, E. Street and K. Jensen for the iodine-thiosulfate titrations, and E. Sherry for the X-ray powder patterns.
9. A saturated aqueous solution of CsCl has a concentration of ~1622 g/L.
10. Power Designs, Inc., New York, NY, Model 2005.
11. R. K. Brown and J. M. Williams, *Inorg. Chem.*, **17**, 2607 (1978).
12. Thermogravimetric analysis performed by Dr. Ted Lance-Gomez of Argonne National Laboratory.
13. J. M. Williams, P. L. Johnson, A. J. Schultz, and C. C. Coffey, *Inorg. Chem.*, **17**, 834 (1978).
14. R. K. Brown, P. L. Johnson, T. J. Lynch, and J. M. Williams, *Acta Crystallogr.*, **B34**, 1965 (1978).
15. R. K. Brown, D. A. Vidusek, and J. M. Williams, *Inorg. Chem.*, **18**, 801 (1979).
16. R. Besinger, D. A. Vidusek, D. P. Gerrity, and J. M. Williams, *Inorg. Synth.*, **20**, 20 (1980).
17. This synthesis is a variation of that reported in 1871. See R. J. Friswell, *Lieb. Ann.*, **159**, 384 (1871).
18. M. A. Beno, F. J. Rotella, J. D. Jorgensen, and J. M. Williams, *Inorg. Chem.*, **20**, 1802 (1981).

Chapter Six

STOICHIOMETRICALLY
UNCOMPLICATED COMPOUNDS

33. TETRAAMMONIUM DIPHOSPHATE

Submitted by RICHARD C. SHERIDAN*
Checked by WILLIAM J. KROENKE†

$$C_3H_6N_6 + H_3PO_4 \longrightarrow C_3H_6N_6 \cdot H_3PO_4$$

$$2C_3H_6N_6 \cdot H_3PO_4 \longrightarrow (C_3H_6N_6)_2 \cdot H_4P_2O_7 + H_2O$$

$$(C_3H_6N_6)_2 \cdot H_4P_2O_7 + 4NH_3 \longrightarrow 2C_3H_6N_6 + (NH_4)_4P_2O_7$$

Ammonium diphosphate cannot be made by thermal dehydration of ammonium phosphates because both water and ammonia are split off, and the resulting product is a mixture of short-chain ammonium polyphosphates and short-chain polyphosphoric acids. Ammonium diphosphate usually is prepared from pure diphosphoric acid.[1] The diphosphoric acid is crystallized from a polyphosphoric acid mixture containing 80% P_2O_5[2] or made by running a solution of sodium diphosphate through an ion exchange column.[3] The resulting diphosphoric acid then is treated with ammonia. Ammonium diphosphate also has been made by heating a mixture of urea and phosphoric acid,[4,5] but by-products such as cyanuric acid are tedious to remove. The present method is based on the reac-

*Division of Chemical Development, National Fertilizer Development Center, Tennessee Valley Authority, Muscle Shoals, AL 35660.
†B. F. Goodrich Company, 9921 Brecksville Road, Brecksville, OH 44141.

tion of melamine (1,3,5-triazine-2,4,6-triamine) and phosphoric acid to form melaminium phosphate, which is converted to melaminium diphosphate by heating at 250°. Treatment of this product with ammonia yields tetraammonium diphosphate.[6]

Procedure

Melamine (126.1 g) is added in small portions to a mechanically stirred mixture of reagent-grade 85% phosphoric acid (146.6 g) and 400 mL water over a 10-minute period. The temperature rises from 25° to about 45°, and the thick slurry of melaminium phosphate is stirred for an additional 20 minutes. The crystals are collected by vacuum filtration and then washed several times with water to remove adhering phosphoric acid. The melaminium phosphate is dried to constant weight in an oven at about 100°. The dried product weighs 220.0 g, or 98% of the theoretical value. It has the refractive indices reported for melaminium phosphate,[7] and chemical analysis shows that it contains 37.3% N and 31.8% P_2O_5 (calculated for $C_3H_6N_6 \cdot H_3PO_4$: N, 37.5; P_2O_5, 31.7). The melaminium phosphate then is heated in a porcelain evaporating dish in an oven at 250° for 1 hour to eliminate water and condense it to melaminium diphosphate (weight loss %: calcd., 4.0; found, 4.1). The product is identified by its refractive indices[7] and by chemical analysis. *Anal.* Calcd. for $(C_3H_6N_6)_2 \cdot H_4P_2O_7$: N, 39.0; P_2O_5, 32.9. Found: N, 38.6; P_2O_5, 32.9.

The melaminium diphosphate (215.1 g) is added in small portions over a 20-minute period to 250 mL water in a 600-mL stainless steel beaker with mechanical stirring while keeping the pH between 8 and 9 by bubbling in gaseous ammonia. The ammoniation is carried out at about 50° to maintain a fluid mixture and to complete the reaction. The thick slurry is stirred for 10 minutes and then filtered to remove the melamine (98% recovery, identified by X-ray powder diffraction[8]). The melamine is washed with 50 mL water and the wash water is added to the filtrate. The combined filtrate and wash are treated dropwise with 250 mL methanol to crystallize the ammonium diphosphate. Mechanical stirring is used, and the temperature of the solution should be about 30-35° to avoid crystallization of the monohydrate which forms at 25° or lower temperatures. The crystals are collected by vacuum filtration and rinsed with methanol. The methanol adhering to the crystals is removed by drawing dry air or nitrogen through the funnel.

The product then is dried in a vacuum desiccator over anhydrous calcium sulfate. (The checker finds that drying for ~64 hours over $CaSO_4$ and 1 hour in vacuo at room temperature produces a 71% yield of product.) The X-ray pattern of the product agrees with that previously reported for tetraammonium diphosphate.[9] Paper chromatography[10] shows that over 95% of the phosphorus is present as diphosphate. The yield is 98.4 g, or 80% based on the melaminium diphosphate.

Anal. Calcd. for $(NH_4)_4P_2O_7$: N, 22.8; P_2O_5, 57.7. Found: N, 22.3; P_2O_5, 57.5.

Properties

Tetraammonium diphosphate loses ammonia slowly and changes to $(NH_4)_3HP_2O_7$ in dry air and to $(NH_4)_3HP_2O_7 \cdot H_2O$ in moist air.[11] In dilute solution, it hydrolyzes by a first-order reaction to phosphate.[12] The rate of hydrolysis is slow in neutral and weakly basic solutions but increases with increasing acidity of the solution. Tetraammonium diphosphate is converted to both anhydrous and hydrated di- or triammonium acid salts by dissolving it in water, adjusting to the proper pH and temperature, and adding ethanol to induce crystallization.[11]

When heated, $(NH_4)_4P_2O_7$ slowly loses ammonia and becomes amorphous.[3] The X-ray powder diffraction patterns,[3,9,11,13] optical properties,[11,13] and infrared absorption spectra[13] for $(NH_4)_4P_2O_7$ and the other ammonium diphosphates have been published. The crystal structure of tetraammonium diphosphate has been described.[14]

References

1. C. Swanson and F. McCollough, *Inorg. Synth.*, **7**, 65 (1963).
2. J. E. Malowan, *Inorg. Synth.*, **3**, 96 (1950).
3. R. V. Coates and G. D. Woodard, *J. Chem. Soc., II*, 1780 (1964).
4. H. A. Rohlfs and H. Schmidt, Ger. Patent 1,216,856 (1966).
5. P. G. Sears and H. L. Vandersall, U.S. Patent 3,645,675 (1972).
6. R. C. Sheridan, U.S. Patent 3,920,796 (1975).
7. S. I. Vol'fkovich, E. E. Zusser, and R. E. Remen, *Bull. Acad. Sci. U.R.S.S., Classe Sci. Chim.*, 571 (1946).
8. L. J. E. Hofer, W. C. Peebles, and E. H. Bean, *U.S. Bur. Mines Bull. No. 613*, 1963, p. 24.
9. *JCPDS X-Ray Powder Diffraction File*, File Card No. 20-102, 1979.
10. T. C. Woodis, Jr., *Anal. Chem.*, **36**, 1682 (1964).
11. A. W. Frazier, J. P. Smith, and J. R. Lehr, *J. Agric. Food Chem.*, **113**, 316 (1965).
12. J. W. Williard, T. D. Farr, and J. D. Hatfield, *J. Chem. Eng. Data*, **20**, 276 (1975).
13. J. R. Lehr, E. H. Brown, A. W. Frazier, J. P. Smith, and R. D. Thrasher, *TVA Chem. Eng. Bull. No. 6*, 1967.
14. N. Middlemiss and C. Calvo, *Can. J. Chem.*, **54**, 2025 (1976).

34.   TETRABUTYLAMMONIUM TETRACHLOROOXO-TECHNETATE(V)

Submitted by ALAN DAVISON,* HARVEY S. TROP,* BRUNO V. DEPAMPHILIS,†
and ALUN G. JONES†
Checked by RUDY W. THOMAS‡ and SILVIA S. JURISSON‡

$$NH_4TcO_4 + 6HCl + Bu_4NCl \xrightarrow[12\,N\ HCl]{} Bu_4N[TcOCl_4] + NH_4Cl + 3H_2O + Cl_2$$

The recent growth of diagnostic nuclear medicine[1] and the use of technetium-99m-containing radiopharmaceuticals as the agents of choice[2] in this field for imaging organ systems in the diagnosis of disease[3] have resulted in a need to adequately investigate the coordination chemistry of technetium. Such studies, using the long-lived technetium-99 ($t_{1/2}$ = 2.12 × 10^5 years, β^- 0.292 MeV), have been hampered by a general lack of information concerning the stability and accessibility of various oxidation states, including the 5+ oxidation state.[2,4-7] Another problem has been the lack of readily available, stable intermediates that can be used to explore reduced states of technetium. The most useful to date have been the salts of the pertechnetate(VII) and hexachlorotechnetate(IV) anions.[2,4-7] However, hexachlorotechnetate(IV) salts have not found much utility in the preparation of technetium(V) complexes,[8] and many technetium(V) complexes are inaccessible from the reduction of pertechnetate(VII) ion.[9]

Recently, the preparation and crystal structure of $[(C_6H_5)_3PNP(C_6H_5)_3]$-$[TcOCl_4]$, a five-coordinate technetium(V) complex have been achieved.[10] The tetrabutylammonium salt, $(Bu_4N)[TcOCl]_4$, can be easily prepared and isolated from the reduction of pertechnetate(VII) ion in 12 N hydrochloric acid in nearly quantitative yield. It is soluble in a variety of polar organic solvents and reacts easily with a variety of ligands to form oxo technetium(V) complexes.[9]

■ **Caution.** *Technetium-99 is a weak β-emitter (0.292 MeV, $T_{1/2}$ = 2.12 × 10^5 years). Therefore, all manipulations should be carried out in a laboratory equipped with a monitored fume hood with Fiberglas trays and plastic-lined absorbent pads used to control potential spills. Personnel should wear disposable lab coats and gloves at all times. Radioactive wastes, both solid and liquid, must be disposed of in special receptacles. Samples sent outside the laboratory must be sealed in glass bottles within a nonbreakable outer container with cotton as a filler, and the glass bottle must be wipetested for contamination prior to ship-*

*Department of Chemistry, Massachusetts Institute of Technology, Cambridge, MA 02139.
†Department of Radiology, Harvard Medical School and Peter Bent Brigham Hospital, Boston, MA 02115.
‡Department of Chemistry, University of Cincinnati, Cincinnati, OH 45221.

ment. All regulations regarding such transportation[11] *must be followed. A standard beta-gamma counter can be used to detect areas of contamination. Normal laboratory glassware is sufficient to stop virtually all the β-emission from the sample; however, concentrated samples of over 0.7 mCi/mL can produce small amounts of Bremstrahlung from the action of the β-particles on the glass.*

Starting Materials

Aqueous solutions of $NH_4[TcO_4]$ (New England Nuclear, Billerica, MA) are standardized easily using ultraviolet spectroscopy[12] (λ_{max} 244 nm; ϵ_{max} 6220 L mole^{-1} cm^{-1}).

Procedure

A 50-mL beaker equipped with a Teflon-jacketed stirbar is filled with 32 mL 12 *N* hydrochloric acid. To the stirred solution, 4.0 mL 0.32 *M* (1.28 mmole) $NH_4[TcO_4]$ in water is added. After 10 minutes, 4.0 mL of a 75% w/w tetrabutylammonium chloride solution (Pfaltz & Bauer) is added dropwise to the green solution. The resultant gray-green microcrystalline powder is filtered and collected on a medium-porosity fritted filter. The solid is washed, and any residual product transferred to the filter with 3.0 mL 12 *N* hydrochloric acid, followed by five 3.0-mL aliquots of isopropyl alcohol or 1-methylethanol. The solid is dried in vacuo for 2 hours. The yield, based on ammonium pertechnetate, is 0.63 g (99%).

Anal. Calcd. for $C_{16}H_{36}Cl_4NOTc$: C, 38.49; H, 7.27; N, 2.81; Cl, 28.40. Found: C, 38.42; H, 7.22; N, 2.67; Cl, 26.34, 29.74.

Properties

Tetrabutylammonium tetrachlorooxotechnetate(V) is a light gray-green solid that dissolves readily in polar organic solvents such as methanol, acetone, dichloromethane, and acetonitrile to yield dark-green solutions when concentrated. Upon addition of water, it disproportionates rapidly to form pertechnetate ion, TcO_4^-, and an insoluble black precipitate, $TcO_2 \cdot xH_2O$. Prolonged contact with concentrated hydrochloric acid results in reduction to the yellow hexachlorotechnetate(IV) ion, $[TcCl_6]^{2-}$. The infrared spectrum, taken as a KBr pellet, exhibits bands due to ν(TcO) at 1019 cm^{-1}(s) and ν(Tc—Cl) at 375 cm^{-1}(m) along with the tetrabutylammonium cation absorptions. The optical spectrum, taken in dichloromethane, has absorption bands at 840 nm (17 L mole^{-1} cm^{-1}), 580 nm (7 L mole^{-1} cm^{-1}), 475 nm (15 L mole^{-1} cm^{-1}), 370 (~100 L mole^{-1} cm^{-1}),* and 293 nm (4450 L mole^{-1} cm^{-1}).

*The effective extinction coefficient of this shoulder is very sensitive to trace impurities in the preparation. Apparent ϵ = 100-140 L mole^{-1} cm^{-1}.

References and Notes

1. S. Baum and R. Bramlet, *Basic Nuclear Medicine*, Appleton-Century-Crofts, New York, 1975.
2. J. A. Siegal and E. Deutsch, in *Annual Reports in Inorganic and General Syntheses—1975*, H. Zimmer and K. Niedenger (Eds.), Academic Press, New York, 1976, p. 311.
3. G. Subramanian, B. A. Rhodes, J. F. Cooper, and V. J. Sodd (eds.), *Radiopharmaceuticals*, The Society of Nuclear Medicine, New York, 1975.
4. R. D. Peacock, in *Comprehensive Inorganic Chemistry*, Vol. 3, J. C. Bailer, H. J. Emeleus, R. S. Nyholm, A. F. Trotman-Dickenson (eds.), Pergamon Press, Oxford, 1973, pp. 877-903.
5. R. D. Peacock and R. D. W. Kemmett, *The Chemistry of Manganese, Technetium, and Rhenium*, Pergamon Press, Elmsford, N.Y., 1975.
6. E. Deutsch and K. Libson, in *Annual Reports in Inorganic and General Syntheses—1976*, H. Zimmer (ed.), Academic Press, New York, 1977, p. 199.
7. (a) R. Münze, *Isotopenpraxis*, **14**, 81 (1978). (b) P. M. Treichel, *J. Organometal. Chem.*, **147**, 239 (1978).
8. A. F. Kuzina, A. A. Oblova, and V. I. Spitsyn, *Russ. J. Inorg. Chem.*, **17**, 1377 (1972).
9. A. Davison, C. Orvig, H. S. Trop, M. S. Sohn, B. V. DePamphilis, and A. G. Jones, *Inorg. Chem.*, **19**, 1988 (1980).
10. F. A. Cotton, A. Davison, V. W. Day, L. D. Gage, and H. S. Trop, *Inorg. Chem.*, **18**, 3024 (1979).
11. Federal Regulations 49CFR173.391-9 and 29CFR11524,11526.
12. G. E. Boyd, *J. Chem. Ed.*, **36**, 3, (1959). Sometimes, NH_4TcO_4 is contaminated with $TcO_2 \cdot xH_2O$, depending upon the age and the supplier. This is most readily removed by centrifuging an aqueous solution prior to standardization.

35. URANIUM(V) FLUORIDES AND ALKOXIDES

Submitted by GORDON W. HALSTEAD* and P. GARY ELLER*
Checked by ROBERT T. PAINE†

Amidst the current resurgence of interest in actinides, the chemistry of uranium in the pentavalent oxidation state has been neglected. This situation is largely due to misconceptions concerning the stability of U(V) with respect to disproportionation and the difficulty in preparing suitable uranium(V) precursors for compound synthesis. Perhaps the most useful precursors for U(V) compound synthesis are uranium pentafluoride and pentaethoxyuranium.[1,2] The standard synthesis of $U(OC_2H_5)_5$, from UCl_4 and sodium ethoxide in the presence of an oxidant, is a tedious procedure which frequently gives poor yields.[2] Several syntheses of uranium pentafluoride have been reported, but each suffers from certain disadvantages.[3] For example, the reactions of UF_4 and F_2,[4] or of HF

*University of California, Los Alamos Scientific Laboratory, Los Alamos, NM 87545.
†Department of Chemistry, The University of New Mexico, Albuquerque, NM 87131.

with UCl_5 or UCl_6,[5] yield UF_5 but require facilities for handling highly corrosive F_2 and HF. Other reported procedures require elevated temperatures (UF_4 + UF_6)[6] inconvenient starting materials, produce mixtures (UF_6 + HBr),[7] or yield only small quantities of product (UF_6 + $SOCl_2$).[8]

   In the following procedures, β-uranium pentafluoride is conveniently prepared by the photochemical reduction of uranium hexafluoride, in a manner similar to an earlier smaller-scale preparation.[9] Pentaethoxyuranium is prepared directly from β-UF_5 and sodium ethoxide in ethanol. The preparation of hexafluorouranium salts from β-UF_5 in nonaqueous solvents is described in a procedure that avoids the use of hydrofluoric acid common to previous methods.[10,11]

■ **Caution.** *Uranium hexafluoride is toxic and is a strong fluorinating agent; it must be handled with care. Carbon monoxide is extremely poisonous.*

A. β-URANIUM PENTAFLUORIDE

$$UF_6 + \frac{1}{2}CO \xrightarrow{h\nu} \beta\text{-}UF_5 + \frac{1}{2}COF_2$$

Procedure (Time Required 1 Day)

A 1-L quartz flask (shown diagrammatically in Fig. 1) (the checker finds that the flasks of this size can be quite fragile. Caution should be exercised in their use) is

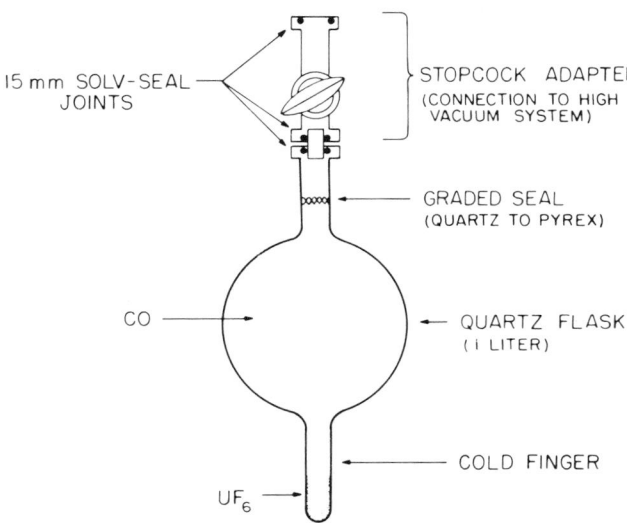

15 mm SOLV-SEAL JOINTS

STOPCOCK ADAPTER
(CONNECTION TO HIGH VACUUM SYSTEM)

GRADED SEAL
(QUARTZ TO PYREX)

CO

QUARTZ FLASK
(i LITER)

COLD FINGER

UF_6

Fig. 1. Reaction vessel for the preparation of β-UF_5.

connected to a greaseless glass or metal high-vacuum line by an attached stopcock adapter. The joints used are greaseless Solv-seal brand available from Fischer & Porter Co., Warminster, PA. The stopcocks are of the greaseless Teflon variety. The flask is evacuated and carefully flamed with a torch under vacuum to remove adsorbed water and then allowed to cool. Approximately 5 g uranium hexafluoride* is vacuum-transferred to the reaction flask by cooling the cold finger with liquid N_2. The liquid N_2 Dewar is then removed, and 600 torr CO is slowly added to the reaction flask while the pressure is observed on a manometer or a Bourdon tube-type gauge. The stopcock is then closed and the flask removed from the vacuum line to an efficient fume hood and placed 10 cm from an Ace Hanovia 550-W UV lamp (No. 6515-36). Irradiation is carried out for 16-18 hours to give a large volume of product in the cold finger and as a coating on the flask walls. The flask is placed back on the vacuum line, and the volatile materials CO, COF_2, and unreacted UF_6 are removed by high vacuum for several hours. The reaction flask is transferred to a drybox, and the fluffy, blue-green β-UF_5 is scraped and shaken into a tared bottle. As many as four identical flasks can be irradiated at one time. With a 5 g charge of UF_6, yields in the range of 70-90% are obtained. If higher charges of UF_6 are used, the percentage yields are lowered substantially due to fogging of the flask windows. In practice, it was found more convenient to use several flasks simultaneously in order to increase the yield of β-UF_5.

Properties

Uranium pentafluoride prepared in the above manner is the low-temperature, β-form, which can be identified by its characteristic X-ray powder pattern.[13] The single-crystal structure has been determined: tetragonal symmetry, space group $I\overline{4}2d(Z = 8)$, with eight-coordination and a geometry intermediate between a dodecahedron and a square antiprism.[14] Uranium pentafluoride is air sensitive and disproportionates in water to give a precipitate of uranium tetrafluoride and a solution of uranyl fluoride.[3] Consequently, UF_5 must be handled in an inert atmosphere. The near-infrared-visible spectrum of uranium pentafluoride in acetonitrile has been determined[1] and is perhaps the most convenient characterization.

---

*Uranium hexafluoride is obtainable from the Department of Energy or Research Organic/Inorganic Chemical Corp. Residual HF, which will attack glass, can be removed by condensation of UF_6 onto dry, powdered KF or NaF[12] or by careful vacuum transfer at $-40°$.

B. PENTAETHOXYURANIUM(V)

$$UF_5 + 5NaOC_2H_5 \longrightarrow U(OC_2H_5)_5 + 5NaF$$

Procedure (Time Required 2 Days)

This preparation must be carried out in an inert atmosphere. Best results are obtained if the ethanol used is absolute grade that has been dried by refluxing over magnesium turnings, distilled, and then degassed prior to use. In a drybox, 15 g UF_5 is added to a 300-mL three-neck flask equipped with a reflux condenser, a nitrogen inlet, and a 250-mL addition funnel. The assembly is then removed from the drybox and connected to a source of dry nitrogen. A solution of 15.3 g freshly prepared $NaOC_2H_5$ in 200 mL dried, degassed ethanol is introduced into the addition funnel via syringe or other inert-atmosphere techniques. The 300-mL flask is cooled in a Dry ice/acetone bath, and the $NaOC_2H_5$ solution is added slowly in portions from the addition funnel. When the addition is complete, the reaction mixture is allowed to warm to room temperature to give a green solution. Refluxing the solution for 4 hours gives a brown solution characteristic of $U(OC_2H_5)_5$.

The solvent is removed in vacuo, yielding a sticky brown residue. The reaction vessel is taken into the drybox and the residue transferred to a short-path vacuum distillation apparatus (good results were obtained with an ACE model 9316 short-path still). The distillation still pot is heated slowly to 170°, and a clear, brown distillate comes over smoothly at a head temperature of 120-130° and a pressure of 10^{-3} torr. The apparatus is cooled to room temperature and taken into a drybox where the liquid $U(OC_2H_5)_5$ is transferred to a tared bottle. Yield 12 g (60%). The yields obtained by the checker ranged 45-50%.

Anal. Calcd. for $UO_5C_{10}H_{25}$; C, 25.92; H, 5.44. Found: C, 25.74; H, 5.34.

Properties

Pentaethoxyuranium is a dark-brown, mobile liquid with a density of 1.71 g/cm^3 at 25°.[2] The liquid boils at 123° at 0.001 torr and is thermally stable below 170°. The compound is readily hydrolyzed and is oxidized by dry oxygen (in the presence of excess $NaOC_2H_5$) to hexaethoxyuranium.[2] Pentaethoxyuranium has a dimeric structure at room temperature as determined by freezing-point depression measurements in benzene.[2] It is miscible with ethanol, diethyl ether, benzene, petroleum ether, ethyl acetate, carbon tetrachloride, chloroform, carbon disulfide, pyridine, dioxane, and nitrobenzene. The proton magnetic resonance spectrum,[15] the near-infrared-visible spectrum,[16] and the C^{13} nmr spectrum[17] of pentaethoxyuranium have been reported and can be used for characterization purposes.

C. HEXAFLUOROURANIUM(V) SALTS

$$UF_5 + MF \longrightarrow M[UF_6]$$

$$(M = Na^+, K^+, [(C_6H_5)_3P]_2N^+)$$

Procedure (time required 4 hours)

1. Sodium Hexafluorouranate(V)   Uranium pentafluoride (2.00 g) and dried NaF (0.25 g) are combined in a 250-mL flask in a drybox. The flask is then attached to a vacuum line, and approximately 150 mL of dry acetonitrile is vacuum transferred into the flask. (The acetonitrile can be dried by distillation from P_4O_{10} or CaH_2.) The mixture is allowed to warm to room temperature. With stirring, the NaF gradually dissolves over a period of 12 hours. The solvent is removed in vacuo, quantitatively yielding $NaUF_6$ in the rhombohedral form (X-ray powder diffraction). The potassium salt is prepared similarly.

2. μ-Nitrido-bis(triphenylphosphorus)(1+) Hexafluorouranate(V) μ-Nitrido-bis-(triphenylphosphorus)(1+) fluoride, [PPN] F, may be prepared by the reported procedure.[18] To a solution of UF_5 (1.35 g) in 75 mL dried, degassed CH_3CN is added under nitrogen a solution of 2.28 g [PPN] F in 50 mL CH_3CN. The solution is stirred for several hours and the solvent is then removed in vacuo until the volume is approximately 20 mL. The resulting light blue-green precipitate is filtered through a coarse-porosity Schlenk frit and vacuum dried to yield 2.40 g (66%) of [PPN] UF_6.

Properties

The hexafluorouranate products are all air-sensitive, pale blue-green powders. The reaction of $[UF_6]^-$ with water produces $[UF_6]^{2-}$ and $[UO_2F_4]^{2-}$ whose presence in the product may be determined by the visible-near-infrared spectrum in acetonitrile.[10] The characteristic fundamental ν_3 mode of $[UF_6]^-$ is found at 525 cm^{-1} in the infrared spectrum of the salts.[11] Reactions of tetraalkylammonium fluorides with uranium pentafluoride in acetonitrile yield the corresponding $[UF_6]^-$ salts, but these salts were found to be unstable under vacuum after removal of acetonitrile. They can be prepared less conveniently in HF, however.[10,11] The uranium in $NaUF_6$ is six-coordinate in solution[1] and in the solid state,[19] whereas in KUF_6 the uranium is six-coordinate in solution and eight-coordinate in the solid state.[1] The exact geometry of the isolated

[UF$_6$]$^-$ anion in the Cs and PPN salts has been determined by single-crystal X-ray studies.[20,21]

Acknowledgments

This work was performed under the auspices of the U.S. Department of Energy.

References

1. G. W. Halstead, P. G. Eller and M. P. Eastman, *Inorg. Chem.*, **18**, 2867 (1979).
2. R. G. Jones, E. Bindschadler, D. Blume, G. Karmas, G. A. Martin, Jr., J. R. Thirtle, F. A. Yeoman, and H. Gilman, *J. Am. Chem. Soc.*, **78**, 4287 (1956). H. Gilman et al., *ibid.*, 4289, 6030.
3. D. Brown, *Halides of the Lanthanides and Actinides*, Wiley-Interscience, New York, 1968, p. 36.
4. L. B. Asprey and R. A. Penneman, *J. Am. Chem. Soc.*, **89**, 172 (1967).
5. J. J. Katz and E. Rabinowitch, *The Chemistry of Uranium*, Vol. 5, National Nuclear Energy Series, Division VIII, McGraw-Hill, New York, 1951, p. 386.
6. A. von Grosse, *U.S. Report TID-5290*, 1958, p. 315.
7. A. S. Wolf, W. E. Hobbs, and K. E. Rapp, *Inorg. Chem*, **4**, 755 (1965).
8. B. Moncela and J. Kikindai, *C. R. Hebd. Seances Acad. Sci., Ser. C*, **267**, 1485 (1968).
9. R. T. Paine and L. B. Asprey, *Inorg. Synth.*, **19**, 137 (1979).
10. J. L. Ryan, *Inorg. Synth.*, **15**, 225 (1974).
11. J. L. Ryan, *J. Inorg. Nucl. Chem.*, **33**, 153 (1971).
12. A. V. Grosse, *The Chemical Properties of UF$_6$*, Report A-83, 1941, pp. 15-16.
13. W. H. Zachariasen, *Acta Crystallogr.*, **2**, 296 (1979).
14. R. R. Ryan, R. A. Penneman, L. B. Asprey, and R. T. Paine, *Acta Crystallogr.*, **B32**, 3311 (1976).
15. D. G. Karraker, T. H. Siddall, and W. E. Stewart, *J. Inorg. Nucl. Chem.*, **31**, 711 (1969).
16. D. G. Karraker, *Inorg. Chem.*, **3**, 1618 (1964).
17. P. G. Eller and P. J. Vergamini, *Inorg. Chem.*, accepted.
18. A. Martinsen and J. Songstad, *Acta Chim. Scand.*, **A31**, 645. (1977).
19. G. D. Sturgeon, R. A. Penneman, F. H. Kruse, and L. B. Asprey, *Inorg. Chem.*, **4**, 748 (1965).
20. A. Rosenzweig and D. T. Cromer, *Acta Crystallogr.*, **23**, 865 (1967).
21. M. P. Eastman, P. G. Eller, and G. W. Halstead, *J. Inorg. Nucl. Chem.*, in press.

36. SODIUM CYANOTRI[(²*H*)HYDRO]BORATE(1-)

Submitted by T. M. LIANG* and M. M. KREEVOY*
Checked by D. GAINES† and G. STEEHLER†

*Department of Chemistry, University of Minnesota, Minneapolis, MN 55455.
†Department of Chemistry, University of Wisconsin, Madison, WI 53706.

$$BH_3(CN)^- + 3\,{}^2H^+ \longrightarrow B\,{}^2H_3(CN)^- + 3H^+$$

Sodium cyanotri[(^{2}H)hydro]borate(1–) (Na[B^{2}H$_3$(CN)]) is useful in the preparation of certain labeled compounds, particularly amines and amino acids prepared by reductive amination.[1] A preparation of Na[B^{2}H$_3$(CN)] by exchange has been described but it is tricky to use, particularly if a substantial amount of Na[B^{2}H$_3$(CN)] is to be prepared, because the pH must be maintained at 2.0 ± 0.2 during the whole reaction period. If it falls too low, there are excessive losses by hydrolysis; and if it is too high, exchange is incomplete. Since Na[BH$_3$(CN)] consumes acid by a competing reaction and also is frequently contaminated with basic impurities, acid must be added continously during the course of the reaction. The present preparation utilizes the same exchange, shown in Eqs. 1 and 2:

$$BH_3CN^- + {}^2H^+ \rightleftharpoons H\,{}^2H\cdot BH_2CN \qquad (1)$$

$$H\,{}^2H\cdot BH_2CN \rightleftharpoons H^+ + BH_2\,{}^2HCN^- \qquad (2)$$

and can be repeated until the desired level of overall deuteration is achieved. Although this procedure has not been used here to prepare the ^{3}H derivative, there appears to be no reason it cannot also be prepared in this way.

The reaction is carried out under conditions giving slower rates of both exchange and hydrolysis,[2] so that the acid concentration is adjusted simply by adding the required amount of acid at the beginning. The ratio of exchange to hydrolysis is more favorable in the present solvent system, so that the yield should be good, even if there are minor deviations from the procedure described.

Procedure

■ **Caution.** *This procedure should be carried out under a well ventilated hood, because some hydrogen cyanide will be liberated by the competing hydrolysis.*

Ten grams (0.16 mole) Na[BH$_3$(CN)] (Alfa Division, Ventron Corp.) was mixed with 80 mL reagent-grade dimethyl sulfoxide (DMSO) and 12 mL ^{2}H$_2$O (Alfa Division, Ventron Corp.; 99.8% deuterated). This mixture was stirred until a homogeneous solution was obtained. It is essential that the next step not be undertaken till a homogeneous solution has been achieved. Undissolved Na[BH$_3$(CN)] will quickly dissolve in and be hydrolyzed by the aqueous acid. The reaction mixture was cooled in an ice bath to 0°. Then, 8 mL 20% ^{2}H$_2$SO$_4$-80% ^{2}H$_2$O was added slowly, from a pressure-relieved dropping funnel protected from atmospheric moisture, the temperature in the reaction mixture being kept below 20°.

■ **Caution.** *Addition of water and/or acid to DMSO is exothermic, so this operation must be carried out carefully, with cooling.*

After the addition was complete, the reaction mixture, which is initially homogeneous, was stirred at room temperature for 2 hours while exchange took place. Then, 5 g anhydrous Na_2CO_3 was added, and the mixture was stirred for 30 minutes to neutralize the acid. This mixture was allowed to settle for 30 minutes and the supernatant liquid was decanted. The solid residue was washed with 10 mL tetrahydrofuran (THF), and the washings were added to the decanted liquid. From the combined liquids, the ^2H_2O and THF were removed by distillation at room temperature. The ^2H_2O, diluted to about 80% deuterium by the exchange, can be recovered for reuse or reenrichment if desired. To the DMSO solution was then added a fresh 12 mL ^2H_2O and 8 mL 20% $^2H_2SO_4$-80% ^2H_2O, as before. Exchange and separation of the product-containing DMSO were carried out as before. Then 3.2 L dry, technical-grade *p*-dioxane was added to the DMSO solution.

The product separated as a white, crystalline *p*-dioxane solvate. The *p*-dioxane can be recovered from the filtrate by distillation. The product was recrystallized from an ethyl acetate/*p*-dioxane mixture, as described by Borch and co-workers.[1] The solid was stirred for 2 hours with 125 mL ethyl acetate. A small amount of residual solid was filtered off, and the ethyl acetate solution was heated to its boiling point. Dry, technical-grade *p*-dioxane (75 mL) was added to the hot solution, which was slowly cooled to room temperature and then chilled. The product separated as a well-crystallized *p*-dioxane solvate. This was dried in vacuo for 2 hours at room temperature, then for 6 hours at 80°. Five grams white, powdery, very hygroscopic product, $Na[B^2H_3(CN)]$, was isolated (0.076 mole, 47% yield). The checkers report 39% yield in a half-scale preparation. They could not detect any residual $BH^2H_2(CN)^-$.

Properties

The ^{11}B nmr spectrum of this product showed a strong singlet, due to $[B^2H_3(CN)]^-$, bisecting a weak doublet, due to $[BH^2H_2(CN)]^{1-}$, and no other visible lines. Integration indicated 95-96 atom percent deuterium in the product.

References

1. R. F. Borch, M. D. Bernstein, and H. D. Durst, *J. Am. Chem. Soc.*, **93**, 2897 (1971).
2. M. M. Kreevoy and E. H. Baughman, *Finn. Chem. Lett.*, 35 (1978).

37.   POTASSIUM PENTAFLUOROOXOMOLYBDATE(V)

Submitted by M. C. CHAKRAVORTI* and S. C. PANDIT*
Checked by T. CHANDLER† and JOHN ENEMARK†

$$MoO(OH)_3 + 5HF \longrightarrow H_2[MoF_5O] + 3H_2O$$

$$H_2[MoF_5O] + 2KHF_2 \longrightarrow K_2[MoF_5O] + 4HF$$

Remarkably little work has been done on the fluoro oxo complexes of molybdenum(V) compared to a very large volume of work done on the analogous chloro oxo complexes. The preparation of the first compound of this type, namely, $K_2[MoF_5O]$, was briefly reported[1] in 1957. Hargreaves and Peacock[1] prepared the compound by fusing $K[MoF_6]$ with potassium(hydrogen difluoride) in a carbon dioxide atmosphere and then extracting the melt with moist acetone. French workers[2,3] have determined the crystal structure of a hydrated complex, namely, $K_2[MoF_5O] \cdot H_2O$. The esr spectra of ammonium, zinc, potassium, and copper(II) salts of $[MoF_5O]^{2-}$ in aqueous hydrofluoric acid solution have been examined.[4] In a recent communication,[5] we have described a very convenient method for the preparation of a large number of salts of the general formulae $M_2[MoF_5O]$ (M = K, Rb, and Cs), $(LH)_2[MoF_5O]$ (L = hydroxylamine, guanidine, 2,6-lutidine, 1-naphthylamine, etc.), and $L'H_2[MoF_5O]$ (L' = ethylenediamine). All these were prepared by crystallizing solutions of $MoO(OH)_3$ and the alkali metal fluorides or the bases in 40% hydrofluoric acid. We describe here a procedure for preparing $K_2[MoF_5O]$ employing $MoO(OH)_3$ which is itself easily prepared[6] from ammonium molybdate in good yields (the procedure is given at the end). The corresponding rubidium and cesium salts can also be prepared by similar methods.

Procedure

■ **Caution.** *Hydrofluoric acid is toxic and all operations with it should be carried out in an efficient fume hood.*

Two grams (12.3 mmole) of $MoO(OH)_3$ is dissolved in a small volume (6 mL) of 40% hydrofluoric acid in a polyethylene beaker. To this, 4.8 g (61.5 mmole) of potassium(hydrogen difluoride) dissolved in 10 mL 40% hydrofluoric acid is added. Fine, green crystals appear on scratching with a polyethylene stirrer. After standing for ½ hour, the product is filtered under suction using a small Büchner

*Department of Chemistry, Indian Institute of Technology, Kharagpur 721 302, India.
†Department of Chemistry, University of Arizona, Tucson, AZ 85721.

funnel the pores and the sides of which are covered well with filter paper. It is poured onto a filter paper and dried scrupulously by pressing between the folds of filter papers. (The moist sample turns blue on standing.) The complex is transferred into a polyethylene dish and dried in a desiccator over concentrated sulfuric acid to constant weight (about 5 days are sufficient). A small dish containing pellets of caustic soda is also placed in the desiccator. Yield is 2.1 g (60%).

Anal. Calcd. for $K_2[MoF_5O]$: Mo, 33.6; F, 33.3; K, 27.4. Found: Mo, 33.9; F, 32.9; K, 27.8; oxidation state of molybdenum, +5.03.

For the analysis of molybdenum, the sample is decomposed by fuming with a few drops of nitric acid and sulfuric acid in a platinum crucible and the molybdenum is determined gravimetrically[7] as the 8-quinolinol complex. From the filtrate, potassium is determined gravimetrically as K_2SO_4. Fluoride is determined by titration with a standard solution of thorium nitrate using sodium alizarinsulfonate as indicator, after steam distillation of fluorosilicic acid.[8] The determination of the oxidation state of molybdenum is carried out by oxidizing a known amount of the compound with a known amount of potassium dichromate in hot 2 N sulfuric acid and titrating the excess dichromate with standard Fe^{2+} solution.

Properties

The compound is sufficiently pure for most uses. It can, however, be recrystallized from 40% hydrofluoric acid. A moist sample turns blue, but the dry sample is stable in air. It is soluble in water, but the green solution gradually changes to red brown. It is paramagnetic[5] (μ_{eff} is 1.55 BM at 30°). The IR spectrum has been studied.[5] The Mo=O bands occur at 980(s), 943sh(w), and 910(m) cm^{-1}. The crystal spectra[9] of $[MoF_5O]^{2-}$ in a host lattice of $K_2[SnF_6] \cdot H_2O$ show bands at 22,000, 21,200, 13,100, and 12,300 cm^{-1}.

References and Notes

1. G. B. Hargreaves and R. D. Peacock, *J. Chem. Soc.*, 4212 (1957).
2. D. Grandjean and R. Weiss, *C. R. Seanc. Acad. Sci. Paris*, **263C**, 58 (1966).
3. D. Grandjean and R. Weiss, *Bull. Soc. Chim. Fr.*, 3040 (1967). *Ibid.*, 3054 (1967).
4. N. S. Garif'yanov, V. N. Fedotov, and N. S. Kucheryavenko, *Izv. Akad. Nauk SSSR, Ser. Khim.*, 743 (1964).
5. M. C. Chakravorti and S. C. Pandit, *J. Coord. Chem.*, 5, 85 (1976).
6. W. G. Palmer, *Experimental Inorganic Chemistry*, University Press, Cambridge, 1954, p. 406. Procedure for trihydroxooxomolybdenum(V), $MoO(OH)_3$ – six g ammonium (para) molybdate, $(NH_4)_6Mo_7O_{24} \cdot 4H_2O$, is dissolved in warm water (25 mL) containing 1 mL concentrated hydrochloric acid. The solution is poured into 75 mL 4.5 N hydrochloric acid. It is transferred to a glass-stoppered bottle and 20 mL of clean mercury is added. The mixture is shaken vigorously for 15 minutes. The dark-red solution is filtered

under suction using a Büchner funnel, and the precipitate is washed with small portions of air-free water (freshly boiled distilled water which is subsequently cooled). The filtrate is transferred to a flask (500 mL capacity) fitted with a gas inlet and outlet. In portions, 22 g powdered ammonium carbonate (NH_4HCO_3 + $NH_2 \cdot COONH_4$ from BDH, laboratory reagent grade) is added. When this has dissolved and reacted, a stream of carbon dioxide gas is passed into the mixture which is heated to boiling. The hydroxide begins to settle after a few minutes. The boiling is continued for 30 minutes and the mixture is allowed to cool, maintaining the stream of carbon dioxide. The brown precipitate is filtered under suction using a Büchner funnel, and washed several times with air-free water and then thoroughly with acetone. It is then transferred to a dish and dried overnight in an evacuated desiccator. The yield is 5.5 g.

7. A. I. Vogel, *A Textbook of Quantitative Inorganic Analysis*, English Language Book Society and Longman, 1973, p. 508.
8. G. Charlot and D. Bezier, *Quantitative Inorganic Analysis*, English translation by R. C. Murray, Methuen & Co. Ltd., London, 1957, p. 424.
9. R. A. D. Wentworth and T. S. Piper, *J. Chem. Phys.*, **41**, 3884 (1964).

38. *CYCLO*-OCTASULFUR MONOXIDE

Submitted by RALF STEUDEL* and TORSTEN SANDOW*
Checked by C. LAU† and T. CHIVERS†

$$(CF_3CO)_2O + H_2O_2 \longrightarrow CF_3CO_3H + CF_3CO_2H$$
$$S_8 + CF_3CO_3H \longrightarrow S_8O + CF_3CO_2H$$

Cyclo-octasulfur monoxide (S_8O) was discovered in 1972 and represented the first example of a new type of sulfur oxides containing homocyclic sulfur rings with one or more exocyclic oxygen atoms attached to the ring. Cyclo-octasulfur monooxide was first prepared in a laborious procedure from thionyl chloride and heptasulfane (H_2S_7) in a carbon disulfide/dimethyl ether mixture at $-40°$ applying the dilution principle.[1] Later, it was found that the compound can be prepared much more conveniently by oxidation of S_8 under mild conditions thus preserving the eight-membered ring.[2] This reaction has since been used for the analogous preparation of S_6O,[3] S_7O,[4] S_7O_2,[5] S_9O,[6,7] and $S_{10}O$[6,7] from the corresponding homocyclic sulfur molecules.

The procedure given allows the preparation of S_8O within three days; it can be scaled down to 1/10 without loss in yield and purity of the product.

*Institut für Anorganische und Analytische Chemie, Technische Universität Berlin, D-1000 Berlin 12, Federal Republic of Germany.
†Department of Chemistry, University of Calgary, Calgary, Alberta, Canada T2N 1N4.

Procedure

Chemicals. Dichloromethane was freshly distilled from P_4O_{10} and passed over a column packed with basic Al_2O_3 (500 g Al_2O_3 per liter CH_2Cl_2) immediately prior to use to remove residual H_2O and ethanol. Pentane and carbon disulfide were dried by distillation from P_4O_{10}. Hydrogen peroxide (80% by weight), a highly corrosive liquid, is commercially available. (■ **Caution.** *Considerable care should be exercised in the use of this oxidant. Avoid contact with reducing agents or metallic parts such as syringes.*) Alternatively, H_2O_2 can be prepared from less concentrated solutions (50%) by carefully controlled partial evaporation in a bulb tube with an oil-pump vacuum. The H_2O_2 concentration can be checked by iodometric titration or, more conveniently, by the density, which is a linear function of the concentration in weight-% (H_2O:1.00, H_2O_2:1.45 g/cm^3 at 25°).

Peroxyacid Solution. In a 500-mL flask, 13.2 mL aqueous H_2O_2 solution (80%; d = 1.36 g/cm^3) is dispersed in 296 mL dry CH_2Cl_2 and cooled to 0°. Within 30 minutes, 88 mL trifluoroacetic acid anhydride (98%) is added dropwise with vigorous stirring and exclusion of moisture (a dropping funnel equipped with pressure equalization tube and fitted with a drying tube is recommended). The mixture is then allowed to warm up to 25°, and stirring is continued for about 15 minutes until no H_2O_2 droplets are discernible; 1 mL of this solution contains 1 mmole CF_3CO_3H, and immediate use is recommended.

Oxidation of S_8. In an 1-L flask, 80 g S_8 (0.312 mole) is dissolved in 300 mL CS_2 at 20° and cooled to 0°. Over a period of 2 hours, 374 mL peroxyacid solution (0.374 mole CF_3CO_3H) is added dropwise with mechanical stirring (magnetic stirrers are too weak) and exclusion of moisture. The mixture turns intensely yellow and becomes turbid with precipitation of S_8O and polymeric sulfur. After stirring an additional 30 minutes at 0°, 250 mL pentane is added dropwise within 1 hour, and the mixture is cooled to −78° for at least 12 hours. The precipitated mixture of S_8, S_8O, polymeric sulfur, and CF_3CO_2H is isolated on a glass frit at 25°, at which temperature CF_3CO_2H (mp − 15°) melts and therefore is removed when the product is washed twice with 50 mL CH_2Cl_2. After drying in an oil-pump vacuum for 10 minutes, 66-70 g crude S_8O is obtained.

Recrystallization of the crude product involves dissolution in 2 L CS_2 at 25°C with stirring for 1 hour, filtration, and cooling the solution to −78° for at least 12 hours. The crystals obtained are isolated on a glass frit at 25°, dried in vacuo, and recrystallized once more from 0.75-1.0 L CS_2. Yield is 5.5-6.0 g (7%), mp 78-79° (dec.).

Properties

S_8O forms orange-yellow crystals which decompose slowly at 25° with evolution of SO_2 but can be stored for weeks in a dry atmosphere at −20°. At the melting point, the oxygen is given off quantitatively as SO_2. The solubility in CS_2 amounts to about 8 g/L at 25° and 3.5 g/L at −25°. Refluxing the solution results in decomposition within a few minutes accompanied by discoloration and precipitation of polymeric sulfur.[1] Infrared and Raman spectra of S_8O have been recorded,[8] and a full vibrational assignment has been published.[9] The SO stretching vibration is found in the IR as a strong band at 1133 cm⁻¹ in CS_2 solution and at 1085 cm⁻¹ in the solid state (KBr disc).

Owing to the thermal and photochemical sensitivity of S_8O, Raman spectra should be recorded at −100° using a red laser line; alternatively, a rotating sample holder can be used at 25°C. Small amounts of S_8 in S_8O can be discovered by determination of the intensity ratio of the Raman lines at 342 and 218 cm⁻¹. For pure S_8O, this ratio amounts to 0.62 (slit width 1.5 cm⁻¹, −100°) but is decreased by admixed S_8 which also exhibits a strong line at 218 but not at 342 cm⁻¹. Alternatively, the purity of S_8O can be checked by iodometric titration.[1] According to an X-ray structural analysis, the S_8O molecules consist of puckered eight-membered rings with exocyclic oxygen atoms in axial positions.[10] With antimony pentachloride, S_8O forms an adduct, $S_8O \cdot SbCl_5$.[11]

References

1. R. Steudel and M. Rebsch, *Angew. Chem. Int. Ed. Engl.*, **11**, 302 (1972). *Z. Anorg. Allg. Chem.*, **413**, 252 (1975).
2. R. Steudel and J. Latte, *Angew. Chem. Int. Ed. Engl.*, **13**, 603 (1974).
3. R. Steudel and J. Steidel, *Angew Chem. Int. Ed. Engl.*, **17**, 134 (1978).
4. R. Steudel and T. Sandow, *Angew. Chem. Int. Ed. Engl.*, **15**, 772 (1976).
5. R. Steudel and T. Sandow, *Angew. Chem. Int. Ed. Engl.*, **17**, 611 (1978).
6. R. Steudel, T. Sandow, and J. Steidel, unpublished.
7. R. Steudel, in *Gmelin Handbuch der Anorganischen Chemie*, 8th ed., *Schwefel*, Vol. 3, Springer, Berlin, 1980, pp. 3-69.
8. R. Steudel and M. Rebsch, *J. Mol. Spectrosc.*, **51**, 334 (1974).
9. R. Steudel and D. F. Eggers, *Spectrochim. Acta*, **31A**, 871 (1975).
10. R. Steudel, P. Luger, H. Bradaczek, and M. Rebsch, *Angew. Chem. Int. Ed. Engl.*, **12**, 423 (1973). P. Luger, H. Bradaczek, R. Steudel, and M. Rebsch, *Chem. Ber.*, **109**, 180 (1976).
11. R. Steudel, T. Sandow, and J. Steidel, *J. Chem. Soc. Chem. Comm.*, 180 (1980).

Chapter Seven

LIGANDS AND REAGENTS

39. SUBSTITUTED TRIARYL PHOSPHINES

Submitted by JOHN E. HOOTS,* THOMAS B. RAUCHFUSS,* and
DEBRA A. WROBLESKI*
Checked by HOWARD C. KNACHEL†

A number of aryl and alkyl phosphines are readily available commercially or are adequately described in this series. On the other hand, phosphines containing reactive functional groups are not generally available as their syntheses are not widely recognized. These preparations often employ conventional organic protecting group methodology[1] as in our synthesis in Section 39-A, or involve unusual reactions[2] as in Section 39-B.

The two phosphines described in this section are of interest to transition metal chemists, as their chelates are of mechanistic[3] and industrial[4] significance. Perhaps more importantly, these compounds represent useful precursors for the syntheses of a variety of other phosphines whose chemistry has yet to be explored.

*School of Chemical Sciences, University of Illinois, Urbana, IL 61801.
†Department of Chemistry, University of Dayton, Dayton, OH 45469.

A. 2-(DIPHENYLPHOSPHINO)BENZALDEHYDE

$$\text{(1)}$$

$$\text{(2)}$$

$$\text{(3)}$$

$$\text{(4)}$$

*Procedure for Reaction 1**

2-Bromobenzaldehyde C 97% (Aldrich Chemical Co.) (25.0 g, 0.131 mole) ethylene glycol (12.50 g, 0.201 mole), and *p*-toluenesulfonic acid monohydrate (0.11 g, 0.58 mmole) are dissolved in 150 mL toluene in a 500-mL, round-bottom flask. The solution is stirred at reflux while the evolved water is collected in a Dean-Stark trap. After water is no longer evolved (ca. 24 hours, the amount of water collected usually exceeds the theoretical quantity), the solution is cooled and washed successively with 50-mL portions of saturated aqueous NaHCO$_3$ and NaCl. The solution is dried over granular K$_2$CO$_3$, concentrated on a rotary evaporator, and distilled at 135-137°, 4 torr. The yield is 27.86 g (93%) of colorless 2-(2-bromophenyl)-1,3-dioxolane.

^{1}H nmr in CDCl$_3$: δ 3.85-4.2 (O—CH$_2$CH$_2$—O, 4H, multiplet); 6.11 (Ar—CH, 1H, singlet); 7.0-8.1 (Ar—H, 4H, multiplet).

*Reaction has been scaled to 5x this size with improved yields.

Procedure for Reactions 2 and 3

An oven-dried, 1-L, three-neck round-bottom flask is equipped with a septum-sealed 125-mL pressure-equalizing addition funnel, a condenser fitted with a nitrogen inlet, and a glass stopper. Next, 2.96 g (0.122 mole) Mg turnings, a Teflon-coated stirring bar, and 200 mL anhydrous tetrahydrofuran (THF) are added to the flask. The addition funnel is charged with 27.8 g (0.121 mole) of the dioxolane. The reaction is initiated with a crystal of I_2 and a few milliliters of the dioxolane solution. The remaining solution is added slowly in 3-mL aliquots from the addition funnel, and the reaction is then heated to reflux for 30 minutes. The clear, brownish solution is cooled in an ice bath while a solution of 22.4 mL (0.121 mole) redistilled chlorodiphenylphosphine (bp 110-120° at 0.1 torr) in 80 mL anhydrous THF is prepared in the addition funnel.

The phosphine is added dropwise to the stirred Grignard reagent while the temperature is maintained at ca. 5°. After the addition is complete, the reaction is heated to reflux for 10-12 hours. The cooled reaction is cautiously quenched with 75 mL 10% aqueous NH_4Cl, and the nonaqueous phase is further extracted with 100 mL saturated aqueous NaCl. The combined aqueous extracts are washed with 100 mL diethyl ether. The combined *non*aqueous extracts are dried over anhydrous Na_2SO_4 and then concentrated to an oily liquid on a rotary evaporator. The crude product can be recrystallized from hot 95% ethanol, followed by cooling to ca. -25° overnight. (The checker recommends rapid filtration with a prechilled glass frit.) The product, 26.39 g (65%), is isolated as colorless crystals, mp 159-161°.

Procedure for Reaction 4

[2-(1,3-Dioxolan-2-yl)phenyl]diphenylphosphine, 26.3 g (78.7 mmole) isolated from the previous reaction is added to 500 mL acetone in a 1-L, round-bottomed flask equipped with a magnetic stirrer and a condenser. *p*-Toluene-sulfonic acid monohydrate, 0.48 g (2.5 mmole), is added and the solution is refluxed for 8 hours. During this time, the yellow color of the final product develops. The course of the reaction may be conveniently monitored by ^{1}H nmr. The warm solution is diluted with 100 mL H_2O, concentrated to ca. 125 mL, and cooled overnight at ca. -25°.

The precipitated yellow crystals are filtered from the cold solution, affording 20.0 g (53% overall for reactions 2-4) crude 2-(diphenylphosphino)benzaldehyde (the checker obtained a yield of 55%). The product may be easily recrystallized from CH_2Cl_2/CH_3OH. The most effective purification is accomplished by vacuum sublimation at 125° and 0.005 torr onto a water-cooled probe.

Anal. Calcd. for $C_{19}H_{15}OP$: C, 78.61; H, 5.21; P, 10.67. Found: C, 78.45; H, 5.38; P, 10.79.

^{1}H nmr: δ 6.8-8.05 (Ar—H, 14H, multiplet); 10.49 (CHO, 1H, doublet, J_{PH} = 4 Hz). The purified product has mp 118-119°.

Properties

The product exists as a yellow solid which is air stable (as are its solutions). The carbonyl absorbs as a doublet in the IR of the compound as its mineral oil mull (ν_{CO} = 1681(s), 1703(s) cm^{-1}). However, in CH_2Cl_2 solution only one peak is observed at 1676 cm^{-1}.

The formyl group smoothly undergoes condensation reactions with a variety of primary mono- and diamines, affording the corresponding iminophosphine chelating agents.[5] These imines can be reduced to the secondary amines with sodium tetrahydroborate(1-). Sodium tetrahydroborate(1-) also cleanly effects the reduction of the phosphine aldehyde itself to the corresponding benzyl alcohol, which exists as an air-stable, low-melting solid.

Using a completely analogous procedure, the corresponding 2-(diphenyl-arsino)benzaldehyde and 2′-(diphenylphosphino)acetophenone can also be prepared. The reactive 2-(dimethylphosphino)benzaldehyde requires a modified procedure for both the Grignard addition and the hydrolysis.

B. 2-(DIPHENYLPHOSPHINO)BENZOIC ACID

$$2Na + PPh_3 \xrightarrow{NH_3} NaPPh_2 + NaPh$$

$$NaPh + NH_3 \longrightarrow NaNH_2 + PhH$$

$$NaNH_2 + o\text{-}ClC_6H_4CO_2H \longrightarrow Na^+ + NH_3 + o\text{-}ClC_6H_4CO^-$$

Procedure

■ **Caution.** *The preparation should be carried out in a well-ventilated fume hood under an atmosphere of argon.*

An oven-dried, 3-L, three-neck, round-bottom flask equipped with a Dry ice condenser, a glass-covered magnetic stirbar, and a gas inlet tube is charged with 1.5 L anhydrous liquid ammonia. Next, Dry ice and acetone are added to the condenser. Sodium metal (15.3 g, 0.667 mole) in ca. 1-g pieces is then added to the stirred ammonia which results in a blue-colored solution. Triphenylphosphine (87.4 g, 0.33 mole) is added in small portions over 30-40 minutes, and the flask is swirled to dissolve any sodium metal on the sides of the flask. After 2½ hours, the red-orange solution of $NaPPh_2$ is treated with 2-chlorobenzoic acid (52.2 g, 0.33 mole) added in small portions over a 30-40 minute period, followed by the addition of 500 mL anhydrous THF via syringe. The reaction mixture is allowed to slowly warm to room temperature overnight under argon. As it warms, the red solid turns golden in color, and the mild evolution of heat is observed.

The residue is dissolved in 1400 mL water and extracted with 400 mL diethyl ether, which is discarded. The aqueous phase is then filtered, acidified to pH 2 with 60 mL concentrated HCl, and extracted with three 200-mL portions of dichloromethane. The dichloromethane solutions are combined, washed with 500 mL water, and evaporated to ca. 125 mL. Methanol, ca. 50 mL, is then added to precipitate the pale-yellow crystalline product (this sometimes requires inducing) which is collected from the cold solution, affording 50 g (49%) of product. The analytically pure sample is obtained by recrystallization of this product from a minimum amount of boiling methanol, mp 175-176°.

Anal. Calcd. for $C_{19}H_{15}O_2P$: C, 74.50; H, 4.94; P, 10.11. Found: C, 74.54; H, 5.02; P, 10.15.

Properties

The air-stable, crystalline compound is soluble in halogenated solvents and diethyl ether but less soluble in other solvents. The infrared spectrum in Nujol mull indicates ν_{OH} at 3200 cm^{-1}(m), and ν_{CO} at 1690 cm^{-1}(s) and at 1275 cm^{-1}(s). This compound can be esterified with diazomethane to the methyl ester, mp 96-97°, which can be used for other syntheses.

Acknowledgment

Acknowledgment is made to the donors of The Petroleum Research Fund, administered by the ACS for support of this research.

References

1. G. P. Schiemenz and H. Kaack, *Justus Liebigs Ann. Chem.*, **9**, 1480 (1973).
2. R. F. Mason and G. R. Wicker, *Ger. Offen.* 2,264,088 (to Shell Oil Co.), *Chem. Abstr.*, **79**, 136457t (1973).
3. T. B. Rauchfuss, *J. Am. Chem. Soc.*, **101**, 1045 (1979).
4. D. M. Singleton, P. W. Glockner, and W. Keim, *Ger. Offen.* 2,159,370 (to Shell Oil Co.), *Chem. Abstr.*, **77**, 89124 (1972).
5. J. C. Jeffrey, T. B. Rauchfuss, and P. A. Tucker, *Inorg. Chem.*, **11**, 3306 (1980).

40. DIMETHYLPHOSPHINE

Submitted by A. TRENKLE* and H. VAHRENKAMP*
Checked by JOHN SVOBODA† and LEO BREWER†

$$2(CH_3)_2PS-PS(CH_3)_2 + 4P(n\text{-}C_4H_9)_3 + 2\,H_2O \longrightarrow$$

$$3(CH_3)_2PH + (CH_3)_2POOH + 4\,SP(n\text{-}C_4H_9)_3$$

The laboratory methods developed for the preparation of dimethylphosphine[1,2] either work only for small quantitites or require some skill of the researcher in order to achieve the reported yields. We have found a sequence of simple reactions which lead to a good yield of dimethylphosphine in a simple apparatus. They are the desulfurization[3] of tetramethyldiphosphine disulfide coupled with the hydrolytic cleavage[4] of the resulting tetramethyldiphosphine accompanied by the high-temperature disproportionation[5] of the dimethylphosphine oxide formed thereby. Although only three-fourths of the Me_2P units supplied by the starting material can be converted to the desired product, the procedure is advantageous because a nearly quantitative yield based on the given reaction can be achieved. The described method can be scaled up without difficulties. The starting materials are either commercially available or easily prepared.

Procedure

■ **Caution.** *Dimethylphosphine is very volatile, poisonous, and spontaneously flammable in air. The reaction should be carried out in a good hood with appropriate precautions.*

The reaction apparatus consists of a 250-mL, round-bottomed flask attached to a 20-cm Vigreux column topped by a 20-cm distillation condenser with thermometer. The condenser is connected to a 25-mL Schlenk flask immersed in an ice bath which is connected by Tygon tubing over a mercury-filled bubbler to the exhaust system. The 250-mL flask contains a 2-cm Teflon-coated magnetic stirring bar. Oil bath heating and stirring are effected by a heating magnetic stirrer. Prior to use, the apparatus is evacuated and filled with nitrogen.

The reaction flask is charged under nitrogen with 15.0 g (80mmole) tetramethyldiphosphine disulfide,[6] 32.3 g (39.0 mL, 160mmole) tributylphosphine, and 1.44 g (80 mmole) distilled water. The mixture is heated to 160-170° under stirring. When the mixture has become homogeneous, the temperature of the oil

*Chemisches Laboratorium der Universität Freiburg, Albertstr. 21, D-7800 Freiburg, Germany.
†Department of Chemistry, University of California, Berkeley, CA 94720.

bath is raised within 2 hours to 220°. Slow gas evolution occurs from the beginning of the heating period. The gas is mainly dimethylphosphine which reaches the condenser with a temperature of 30-50° and is collected in the ice-cooled receiving flask. The reaction is finished when no more gas is evolved.

After the reaction, the pressure in the apparatus is equalized through the nitrogen inlet. After cooling to room temperature, the apparatus is disconnected with caution because residues may inflame. The receiving flask is kept under nitrogen and cold because of the volatility and inflammability of the dimethylphosphine. The yield is 5.4-6.5 g (72-87% based on the given reaction). The checkers report a somewhat lower yield (~50%) using one-half the above scale.

The yield of dimethylphosphine can be increased by increasing the amount of tributylphosphine which allows use of water in the less volatile form of a hydrate. Thus using the same apparatus, procedure and with the same amount of tetramethyldiphosphine disulfide, using 50.5 g (250mmole) $P(n\text{-}C_4H_9)_3$ and 15.2 g (80mmole) p-toluenesulfonic acid hydrate, the yield of dimethylphosphine is 7.1 g (95%).

Properties

The properties of dimethylphosphine have been reported.[1] It can be stored in the pure state in a deep freezer. It is however more convenient to store and use it in the form of 1- to 3-M solutions in hydrocarbon solvents.

References

1. W. L. Jolly, *Inorg. Synth.*, **11**, 126 (1968).
2. G. W. Parshall, *Inorg. Synth.*, **11**, 157 (1968).
3. L. Maier, *J. Inorg. Nucl. Chem.*, **24**, 275 (1962).
4. L. Maier, in *Organic Phosphorus Compounds*, Vol. 1, G. M. Kosolapoff and L. Maier (eds.), Wiley, New York, 1972, p. 321.
5. H. R. Hays, *J. Org. Chem.*, **33**, 3690 (1968).
6. S. A. Butter and J. Chatt, *Inorg. Synth.*, **15**, 185 (1974).

41. 1,2,3,4,5-PENTAMETHYLCYCLOPENTADIENE

Submitted by JUAN M. MANRIQUEZ,* PAUL J. FAGAN,* LARRY D. SCHERTZ,* and TOBIN J. MARKS*
Checked by JOHN BERCAW† and NANCY MCGRADY†

*Department of Chemistry, Northwestern University, Evanston, IL 60201.
†Department of Chemistry, California Institute of Technology, Pasadena, CA 91125.

The 1,2,3,4,5-pentamethylcyclopentadienyl ligand is a complexing agent of great utility in main-group,[1,2] transition-metal,[3-5] and f-element[6] organometallic chemistry. In comparison to the unsubstituted cyclopentadienyl ligand, it is a greater donor of electron density[3,4] while imparting considerably enhanced solubility and crystallizability to the metal complexes it forms. In several cases, peralkylcyclopentadienyl incorporation also increases the thermal stability of metal-to-carbon sigma-bonded derivatives.[4,7] The latter characteristic appears to reflect the well-known instability of the unsubstituted η^5-C_5H_5 moiety with respect to ring hydrogen atom transfer.[4,6d] The steric influence of the penta-methylcyclopentadienyl ligand is also important, and for actinides it leads to coordinative unsaturation by restricting the number of cyclopentadienyl groups that can be bound to the metal ion.[6]

The first detailed study of pentamethylcyclopentadienyl organometallic chemistry was carried out by King and Bisnette.[3] Pentamethylcyclopentadiene was prepared by the laborious (6 steps) and expensive procedure of deVries.[8] This method was later improved somewhat by substituting CrO_3/pyridine for MnO_2 in the step involving oxidation of 3,5-dimethyl-2,5-heptadien-4-ol (di-*sec*-2-butenylcarbinol).[4a] More recently, Burger, Delay, and Mazenod[9] reported two alternative syntheses of pentamethylcyclopentadiene. The shortest approach (3 steps) begins with expensive hexamethyl Dewar benzene, while the second procedure involves 5 steps and proceeds in rather low overall yield. A three-step preparation by Feitler and Whitesides[10] is more efficient than any of the above methods. Even simpler and more straightforward is the adaptation of the procedure of Sorensen et al.[4c] by Threlkel and Bercaw,[4d] which begins with a commercially available mixture of *cis*- and *trans*-2-bromo-2-butene. The present approach (shown below) is based upon the latter procedure but intro-duces several important modifications which result both in increased economy and the possibility of larger-scale syntheses.

Thus, reduced quantities of solvents, different addition procedures for reagents, and streamlined work-up methods are employed.

Procedure

■ **Caution.** *2-Bromo-2-butene and* p-*toluenesulfonic acid are considered hazardous. Avoid skin contact and inhalation.*

A mixture of *cis-* and *trans-*2-bromo-2-butene was prepared from *cis-* and *trans-*2-butene (Matheson Co., East Rutherford, NJ) by the literature procedure.[11] (Alternatively, this mixture can be purchased from Pfaltz and Bauer, Inc., Stanford, CT; for large-scale syntheses, it is more economical to prepare it.) A 5-L three-necked reaction flask equipped with a 500-mL pressure-equalizing addition funnel, reflux condenser, efficient magnetic stirring bar, and gas inlet is heated with a hot air gun while flushing with argon. (For further scale-up, a mechanical stirrer is recommended.) Next, 58.0 g (8.36 moles) of 3.2-mm-diameter Li(0.02% Na) wire (Alfa Division, Ventron Corp., Danvers, MA) is cut into the flask in ca. 5-mm pieces under a flush of argon. Then, 1600 mL diethyl ether (freshly distilled from Na/K benzophenone) is added to the flask under argon. The composition of the sodium-potassium alloy is ca. 3:1 K/Na by weight. The addition funnel is charged with 120 g (0.88 mole, 90.4 mL) 2-bromo-2-butene, which has been dried over Davison 4 Å molecular sieves. A 10-mL aliquot of 2-bromo-2-butene is added to the lithium/ether mixture with stirring. As the reaction begins,* cloudiness due to suspended LiBr should be observable. The 2-bromo-2-butene is added dropwise at a rate sufficient to maintain reflux of the diethyl ether.

After the addition of the 2-bromo-2-butene is complete (ca. 1.5 hour), a mixture of 166 g (1.88 mole, 184 mL) of ethyl acetate (dried over Davison 4 Å molecular sieves) and 430 g (3.18 mole, 324 mL) 2-bromo-2-butene is added dropwise over a period of 4-5 hours, maintaining reflux. Since the reaction rate often slows markedly midway through the addition, the addition rate should be increased if necessary and then slowed once the reflux becomes vigorous again. After the addition of this mixture is complete, stirring is continued for 15 minutes. Next, the addition funnel is charged with a further 50 g (55.4 mL) dry ethyl acetate, and only a portion of this reagent (usually 10-20 g) is added dropwise to the reaction mixture until refluxing of the ether ceases. No further ethyl acetate should be added. After the reaction mixture has cooled for 4 hours, 1200 mL of a saturated NH_4Cl solution is added dropwise to hydrolyze the remaining lithium. The diethyl ether layer is isolated using a separatory

---

*Experience has shown that rigorously anhydrous and anaerobic conditions are necessary for the success of this reaction. All reagents must be dried appropriately and degassed.

funnel, and the aqueous layer is extracted with three 200-mL portions of diethyl ether. The diethyl ether solutions are combined and concentrated to ca. 350 mL on a rotary evaporator.

The diethyl ether concentrate is added to a slurry of 36 g *p*-toluenesulfonic acid monohydrate (Aldrich Chemical Co., Milwaukee, WI) in 500 mL diethyl ether contained in a 2-L, three-neck flask equipped with a reflux condenser and magnetic stirring bar. The rate of addition should be sufficiently slow that the diethyl ether gently refluxes. The reaction mixture is stirred for 5 minutes after refluxing ceases and is then poured into 1200 mL of a saturated $NaHCO_3$ solution containing 19 g Na_2CO_3. The yellow aqueous phase is removed and extracted with three 200-mL portions of diethyl ether. The combined diethyl ether solutions are dried over Na_2SO_4 and then concentrated to ca. 250-300 mL on a rotary evaporator. The crude product is trap-to-trap distilled in vacuo (with the aid of a warm water bath (35-40°)), and the resulting yellow liquid (85% pure by gas chromatography) is then vacuum distilled under N_2 using a 50-cm Vigreux column. The fraction boiling from 65-70°/20 torr is collected to yield 142 g (53% overall yield based upon ethyl acetate) 1,2,3,4,5-pentamethylcyclopentadiene as a pale-yellow liquid (92% pure by GC). An additional fraction boiling at 70-75°/20 torr can also be collected (85% pure by GC) and represents an additional 15 g (5%) of product.*

Properties

1,2,3,4,5-Pentamethylcyclopentadiene is a colorless to pale-yellow liquid with a sweet olefinic odor. It should be stored under nitrogen in a freezer. The ^1H nmr spectrum (CCl_4) exhibits a multiplet at $\delta = 2.4$ (1H), a broadened singlet at $\delta = 1.75$ (12H), and a sharp doublet at $\delta = 0.95$ (3H, $^3J_{H-H} = 8$ Hz). The infrared spectrum (neat liquid) exhibits significant transitions at 2960 (vs), 2915 (vs), 2855 (vs), 2735 (w), 1660 (m), 1640 (w), 1390 (s), 1355 (m), 1150 (w), 1105 (mw), 1048 (w), 840 (mw), and 668 (w) cm^{-1}. For the preparation of metal complexes, pentamethylcyclopentadiene can be converted to the lithium reagent by treatment with lithium alkyls[4,10] or to the Grignard reagent by refluxing with isopropyl magnesium chloride in toluene.[6c]

Acknowledgments

This research was supported by NSF Grants CHE76-84494A01 and CHE8009060.

---

*The checkers report three fractions selected in this sequence: 46 g (67% purity), 129 g (88% purity), and 12 g (74% purity).

References

1. A. Davison and P. E. Rakita, *Inorg. Chem.*, **9**, 289 (1970).
2. P. Jutzi, F. Kohl, and C. Krüger, *Angew. Chem. Int. Ed.*, **18**, 59 (1979).
3. R. B. King and M. B. Bisnette, *J. Organometal. Chem.*, **8**, 287 (1967).
4. (a) J. E. Bercaw, R. H. Marvich, L. G. Bell, and H. H. Brintzinger, *J. Am. Chem. Soc.*, **94**, 1219 (1972). (b) J. E. Bercaw, *J. Amer. Chem. Soc.*, **96**, 5087 (1974). (c) P. H. Campbell, N. W. K. Chiu, K. Deugan, I. J. Miller, and T. S. Sorensen, *J. Am. Chem. Soc.*, **91**, 6404 (1969). (d) R. S. Threlkel and J. E. Bercaw, *J. Organometal. Chem.*, **136**, 1 (1977), and references therein.
5. (a) P. M. Maitlis, *Acc. Chem. Res.*, **11**, 301 (1978), and references therein. (b) D. P. Freyberg, J. L. Robbins, K. N. Raymond, and J. C. Smart, *J. Am. Chem. Soc.*, **101**, 892 (1979). (c) S. J. McLain, R. R. Schrock, P. R. Sharp, M. R. Churchill, and W. J. Youngs, *J. Am. Chem. Soc.*, **101**, 263 (1979).
6. (a) P. J. Fagan, E. A. Maatta, and T. J. Marks, *ACS Symp. Series*, **152**, 52 (1981). (b) P. J. Fagan, J. M. Manriquez, S. H. Vollmer, C. S. Day, V. W. Day, and T. J. Marks, *J. Am. Chem. Soc.*, in press. (c) P. J. Fagan, J. M. Manriquez, E. A. Maatta, A. M. Seyam, and T. J. Marks, *J. Am. Chem. Soc.*, **103**, 6650 (1981) and references therein. (d) J. M. Manriquez, P. J. Fagan, T. J. Marks, C. S. Day, and V. W. Day, *J. Am. Chem. Soc.*, **100**, 7112 (1978). (e) P. J. Fagan, J. M. Manriquez, and T. J. Marks, in *Organometallics of the f-Elements*, T. J. Marks and R. D. Fischer (eds.), Reidel Publishing Co., Dordrecht, Holland, 1979, Chap. 4.
7. F. H. Köhler, W. Prössdorf, U. Schubert, and D. Neugebauer, *Angew. Chim. Int. Ed.*, **17**, 850 (1978).
8. L. deVries, *J. Org. Chem.*, **25**, 1838 (1960).
9. U. Burger, A. Delay, and F. Mazenod, *Helv. Chim. Acta*, **57**, 2106 (1974).
10. D. Feitler and G. M. Whitesides, *Inorg. Chem.*, **15**, 466 (1976).
11. F. G. Bordwell and P. S. Landis, *J. Am. Chem. Soc.*, **79**, 1593 (1957).

42. VANADIUM DICHLORIDE SOLUTION

Submitted by MARTIN POMERANTZ,* GERALD L. COMBS JR.,* and N. L. DASSANAYAKE*
Checked by GEORGE OLAH† and G. K. SURYA PRAKASH†

$$V_2O_5 + 3Zn + 10HCl \longrightarrow 2VCl_2 + 3ZnCl_2 + 5H_2O$$

Because of the increased use of, and interest in, VCl_2 as a 1-electron reducing agent for a large variety of organic molecules,[1] a short, efficient synthesis of a suitable VCl_2 solution is highly desirable. The current syntheses of VCl_2 include

*Department of Chemistry, University of Texas at Arlington, Arlington, TX 76019.
†Department of Chemistry, University of Southern California, Los Angeles, CA 90007.

the preparation of solid, anhydrous VCl_2 by reduction of VCl_3 at $675°$ in a hydrogen atmosphere[2] and the preparation of solid $VCl_2 \cdot 2THF$ (THF = tetrahydrofuran) which takes more than 12 hours and involves preparing $VCl_3 \cdot 3THF$ from VCl_3 and then reducing this with Zn.[3] For most 1-electron reductions, all that is needed is an aqueous solution of VCl_2, but the procedure of Conant, which requires a 2-day reduction of V_2O_5 with zinc amalgam in HCl,[4] is much too long to be of optimum usefulness.

We now describe a rapid (about 0.5 hour) zinc reduction of the relatively inexpensive V_2O_5 which produces an aqueous solution of VCl_2 which is about 1 N and is suitable directly for a large variety of reductions. The following are representative 1-electron reductions which require such a solution of VCl_2 and an organic solvent which is miscible with water (such as tetrahydrofuran, ethanol, acetone, or acetonitrile) for the organic substrate: bromoketone reduction,[5] preparation of vanadocene type molecules,[3,5] tropyl cation dimerization,[7,8] the reductive dimerization of alcohols,[9,10] the reductive dimerization of α,β-unsaturated ketones and aldehydes and aromatic aldehydes,[4] the reduction of aryl azides,[11] the reduction of sulfoxides,[12] the reduction of benzylic and allylic halides and vic-dihalides,[13] the conversion of quinones to hydroquinones,[14] and the reductive cleavage of oximes.[15]

Procedure

To 11 g Zn powder (0.17 g-atom; Fisher Scientific, Alfa Products) in a three-neck flask equipped with an addition funnel, a condenser with a drying tube, and a nitrogen inlet, 5.7 g (31 mmole) solid V_2O_5 (J. T. Baker, Alfa Products) was added. To this was added dropwise, with stirring, under oxygen-free nitrogen, 60 mL 15% HCl (25 mL of concentrated HCl diluted to 60 mL with distilled water). After stirring at room temperature for 0.5 hour, the solution was filtered to produce a purple VCl_2 solution which was $\sim$1 N (titrated against standard $KMnO_4$ or ferric ammonium sulfate).[16]

Properties

Although the solution of VCl_2 contains considerable $ZnCl_2$, it is suitable for a large variety of reductions. The VCl_2 solution is air sensitive, and for best results should be prepared, filtered, and reacted under an oxygen-free nitrogen or argon atmosphere. We have found, however, that, since many reductions by VCl_2 are quite rapid, we could use the solution immediately after preparation without a nitrogen atmosphere without any significant change in the yield of product. Titration immediately after preparation with standardized $KMnO_4$ solution or ferric ammonium sulfate[16] showed the VCl_2 to be about 1 N (typical values were between 0.89 and 0.98 N). After standing 20 hours under an argon atmosphere, a solution which was originally $0.90\,N$ ($KMnO_4$ titration) became $0.63\,N$. Thus, if handled properly, the VCl_2 solution may be used up to a few hours after preparation with only a small loss.

Acknowledgments

The authors wish to thank the Robert A. Welch Foundation and the Organized Research Fund of UTA for support of this work.

References

1. See, for example, T.-L. Ho, *Synthesis*, 1 (1979).
2. R. C. Yound and M. E. Smith, *Inorg. Synth.*, **4**, 126 (1953).
3. F. H. Köhler and W. Prössdorf, *Z. Naturforsch.*, **32B**, 1026 (1977).
4. J. B. Conant and H. B. Cutter, *J. Am. Chem. Soc.*, **48**, 1016 (1926).
5. T.-L. Ho and G. A. Olah, *Synthesis*, 807 (1976).
6. F. H. Köhler and W. Prössdorf, *Chem. Ber.*, **111**, 3464 (1978).
7. G. A. Olah and T.-L. Ho, *Synthesis*, 798 (1976).
8. M. Pomerantz, G. L. Combs, Jr., and R. Fink, *J. Org. Chem.* **45**, 143 (1980).
9. J. B. Conant et. al., *J. Am. Chem. Soc.*, **47**, 572, 1959 (1925).
10. L. H. Slaugh and J. H. Raley, *Tetrahedron*, **20**, 1005 (1964).
11. T.-L. Ho, M. Henninger, and G. A. Olah, *Synthesis*, 815 (1976).
12. G. A. Olah, G. K. Surya Prakash, and T.-L. Ho, *Synthesis*, 810 (1976).
13. T. A. Cooper, *J. Am. Chem. Soc.*, **95**, 4158 (1973).
14. T.-L. Ho and G. A. Olah, *Synthesis*, 815 (1976).
15. G. A. Olah, M. Arvanaghi, and G. K. Surya Prakash, *Synthesis*, 220 (1980).
16. A. I. Vogel, *A Textbook of Qualitative Inorganic Analysis*, 3rd ed., Longman's, London, 1961, p. 399.

43. URANIUM(IV) CHLORIDE FOR ORGANOMETALLIC SYNTHESIS

Submitted by I. A. KHAN* and H. S. AHUJA*
Checked by K. W. BAGNALL[†] and LEE SINF[†]

$$UO_2 + CCl_4 \longrightarrow UCl_4 + CO_2$$

The synthesis of compounds such as bis (η^8-cyclooctatetraene)uranium [uranocene,[1,2] $U(C_8H_8)_2$] with high volatility has generated renewed interest in organometallic compounds of tetravalent uranium. Uranium(IV) chloride, which

*Chemistry Division, Bhabha Atomic Research Centre, Trombay, Bombay 400 085.
[†]The University of Manchester, Department of Chemistry, Manchester, M13 9PL England.

is the primary material for the synthesis of such compounds, has been prepared by liquid-phase chlorination of UO_3 with hexachloropropene,[3] direct chlorination of uranium hydride,[4] vapor-phase chlorination of UO_2 with carbon tetrachloride,[5] and reacting UF_4 with $AlCl_3$ or BCl_3.[4]

If a high-purity product is not required, the procedure described here is an improvement over the earlier method,[5] since the tedious operation to completely remove water (formed during the reduction of UO_3) from the apparatus is avoided. Active UO_2 is obtained from anhydrous uranyl oxalate. The use of hydrogen gas during chlorination gives a clean reaction. These procedures give UCl_4 that is sufficiently pure for most subsequent synthetic work.

Procedure

■ **Caution.** *The procedure is to be carried out inside a fumehood in a well-ventilated room. Safety precautions in handling hydrogen gas and small amounts of highly toxic CO and $COCl_2$ reported to form during chlorination[6] should be observed.*

The apparatus is shown in Fig. 1. While the apparatus is being purged with oxygen-free nitrogen,[7] the reaction tube is heated to about 373 K and the area of the assembly outside the furnace zone is occasionally flamed to obtain a dry and oxygen-free system. After about 15 minutes, the flow of nitrogen is in-

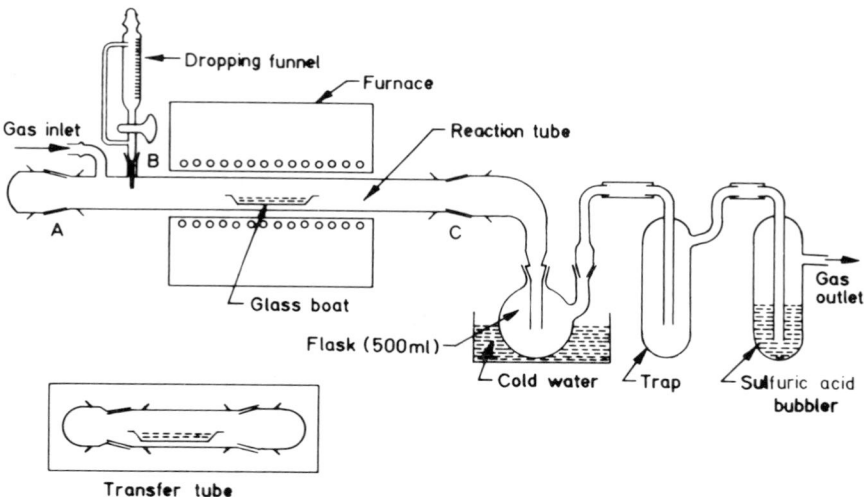

Fig. 1. *Apparatus for the preparation of uranium(IV) chloride: A. Furnace (5 cm dia. × 30 cm long). B. Reaction tube (4.5 cm dia. × 48 cm long) with B40 joints. C. Boat (2.8 cm dia. × 15 cm long).*

creased and the glass boat containing 20 g anhydrous uranyl oxalate* is introduced into the system with the help of a transfer tube (previously flushed with nitrogen) by manipulating the stopper A. Then, the stopper at B is removed and a graduated funnel containing 50 mL A.R. grade CCl_4 (dried and degassed) is connected to the assembly. The nitrogen flushing is continued for another 10 minutes, and this is then replaced by dry hydrogen (deoxygenated using Deoxo catalyst) with a flow rate of about 30 mL/min. After a period of another 10 minutes, the furnace temperature is gradually increased to about 623 K, where uranyl oxalate decomposes. This is indicated by an increase in the flow of gases through the bubbler. The decomposition is complete in less than 5 minutes. The furnace temperature is then raised to 673 K and kept at this temperature for 30 minutes to ensure complete conversion of the oxalate to UO_2.

Carbon tetrachloride is now dropped into the reaction tube at the rate of about 20 mL/hr. *When all the CCl_4 has been added, the furnace is switched off.* During the course of the reaction and also while the assembly is being cooled, the reaction tube outside the furnace is occasionally heated using a hot-air blower to vaporize condensed CCl_4. When the reaction tube is at about 373 K, hydrogen gas is replaced by nitrogen and cooling is continued to room temperature. By maintaining an increased flow of nitrogen, the parts of the assembly beyond C are removed and the transfer tube already purged with nitrogen is introduced at this point. The stopper at A is quickly removed, and the boat containing UCl_4 is pushed into the transfer tube, which is then quickly removed from the assembly and stoppered. The uranium tetrachloride is transferred to a suitable ampul inside a nitrogen-filled drybox and sealed. The UCl_4 weighs 20 g, corresponding nearly to the theoretical yield.†

Anal. Calcd. for UCl_4: U, 62.67; Cl, 37.33. Found: U, 62.81; Cl, 37.27.

Uranium(IV) chloride prepared by this procedure contains traces of free carbon in the range 25-300 ppm. This impurity, however, has no effect for subsequent use in the synthesis of organouranium(IV) and other compounds.

Properties

The uranium(IV) chloride prepared by this method is dark green in color. Since it is sensitive to moisture, it should be handled in a drybox. It melts at 863 K and dissolves readily in water with decomposition. The chloride is soluble in most polar organic solvents but is insoluble in hydrocarbons and diethyl ether. Physical properties and thermochemical data for this compound have been reported.[9]

*$(UO_2)(C_2O_4)\cdot 3H_2O$ is prepared by the reported procedure.[8] It is heated at 423 K under vacuum to give anhydrous $(UO_2)(C_2O_4)$. To avoid any subsequent moisture absorption, this compound is handled in a nitrogen-filled drybox.

†The checkers find the product to be ~98.1% UCl_4. It is contaminated by small amounts of carbon.

References

1. A. D. Streitweiser, Jr., and U. Muller-Westerhoff, *J. Am. Chem. Soc.*, **90**, 7364 (1968).
2. B. Kanellakapulos and K. W. Bagnall, MTP *Int. Rev. Sci. Inorg. Chem.* Vol. 7, University Park Press, Baltimore, 1971.
3. J. A. Hermann and J. F. Suttle, *Inorg. Synth.*, **5**, 143 (1957).
4. M. Oxley, in *Nouveau Traite de Chimie Minerale*, Vol. 15, P. Pascal (ed.), Masson et Cie, Paris, 1961, p. 136.
5. P. W. Wilson, *Synth. Inorg. Metalorg. Chem.*, **3**, 381 (1973).
6. I. V. Budayer and A. N. Volsky, *Proc. U.N. 2nd Int. Conf. Peaceful Uses Atomic Energy, Geneva*, **28**, 316 (1958).
7. D. F. Shriver, *The Manipulation of Air Sensitive Compounds*, McGraw-Hill Book Co., New York, 1969, p. 190.
8. M. G. Brauer, *Hand Book of Preparative Inorganic Chemistry*, Vol. 2, Academic Press, New York, 1965, p. 1442.
9. D. Brown, *Halides of Lanthanides and Actinides*, John Wiley, London, 1968, p. 134.

CORRECTION

PREPARATION OF $[Pt(H_2O)_4]^{2+}$ STOCK SOLUTION

Suggested by SHARON A. JONES,* WILLIAM H. TAYLOR,* MYRL D. HOLIDA,* LORI J. SUTTON,* and JACK M. WILLIAMS†
Checked by A. R. SIEDLE‡

■ **Caution.** *Perchlorates are potentially explosive and must be treated with care.*

The preparation of various platinum(II) complexes from a stock solution of $[Pt(H_2O)_4]^{2+}$ has been described.[1] Unfortunately, this procedure leads to product contamination by mercury(II). Using an alternative procedure developed by Elding,[2] it is possible to prepare a $[Pt(H_2O)_4]^{2+}$ stock solution that avoids this problem.

The procedure of Elding should be consulted. In it, 0.20 M silver(I) perchlorate (400 ml, 1 M HClO$_4$) is added to remove chloride[3] from a solution of K_2PtCl_4 (3.5g) in 1 N HClO$_4$ (400 ml). This is done under argon, in the dark, at

*Research participants sponsored by the Argonne Division of Educational Programs: Sharon A. Jones from Thiel College, Greenville, PA; William H. Taylor from Dartmouth College, Hanover, NH; Myrl D. Holida from Dakota State College, Madison, SD; and Lori J. Sutton from Marywood College, Scranton, PA.
†Correspondent: Chemistry Division, Argonne National Laboratory, Argonne, IL 60439.
‡3M Central Research Laboratory, St. Paul, MN 55101.

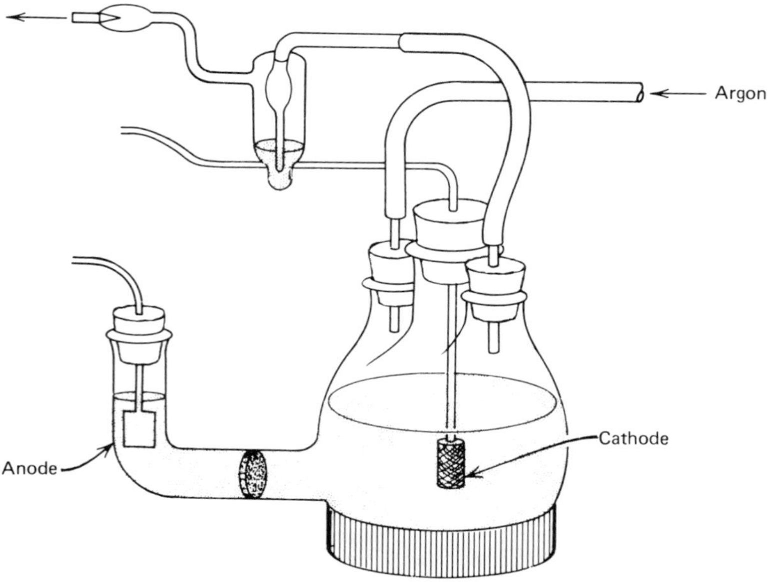

Fig. 1. Electrochemical cell for the removal of residual silver ion. The cathode is a cylindrical platinum net electrode (18 × 40 mm) and the anode is a 20 × 40 mm platinum sheet. A Kimble fine frit (porosity 4-5.5) separates the anode from the cathode compartment. The checker has added a vent to the anode compartment to remove H_2. A self-regulating power supply such as the Tektronix PS 501-2 works well.

70°. The excess silver ion is then removed electrochemically at a cylindrical platinum net electrode (1.37 ± 0.01 V)[4] in an apparatus‡* such as presented in Fig. 1. In a typical run, the electrolysis requires ~110-120 hours. During this time, the current initially is ~6-8 mA, rises to ~9-12 mA after ~60 hours, and drops to ~3-5 mA after 110 hours. The resulting yellow $[Pt(H_2O)_4](ClO_4)_2$ stock solution[5] has been demonstrated to produce bis[2,4-pentanedionato(1−)]-platinum(II) and other platinum(II) complexes uncontaminated by other metal ions.

Acknowledgments

This work was performed under the auspices of the Office of Basic Energy Sciences, Division of Material Sciences, U.S. Department of Energy.

‡*The anode compartment should be vented to allow escape of the H_2 which is formed.

References and Notes

1. S. Okeya ans S. Kawaguchi, *Inorg. Synth.*, **20**, 65 (1980).
2. L. I. Elding, *Inorg. Chem. Acta*, **20**, 65 (1976).
3. The $AgClO_4$ solution must be added slowly over a period of about 7 hours at 70° to avoid precipitation of Ag_2PtCl_4. Stirring is continued at 70° for 5-20 days.
4. This is the voltage measured across the electrodes.
5. The solution should be stored in the dark under argon at ~0°.

INDEX OF CONTRIBUTORS

SUBJECT INDEX

Names used in this Subject Index for Volumes 21–25 are based upon IUPAC *Nomenclature of Inorganic Chemistry*, Second Edition (1970), Butterworths, London; IUPAC *Nomenclature of Organic Chemistry*, Sections A, B, C, D, E, F, and H (1979), Pergamon Press, Oxford, U.K.; and the Chemical Abstracts Service *Chemical Substance Name Selection Manual* (1978), Columbus, Ohio. For compounds whose nomenclature is not adequately treated in the above references, American Chemical Society journal editorial practices are followed as applicable.

Inverted forms of the chemical names (parent index headings) are used for most entries in the alphabetically ordered index. Organic names are listed at the "parent" based on Rule C-10, Nomenclature of Organic Chemistry, 1979 Edition. Coordination compounds, salts and ions are listed once at each metal or central atom "parent" index heading. Simple salts and binary compounds are entered in the usual uninverted way, e.g., *Sulfur oxide* (S_8O), *Uranium(IV) chloride* (UCl_4).

All ligands receive a separate subject entry, e.g., *2,4-Pentanedione*, iron complex. The headings *Ammines, Carbonyl complexes, Hydride complexes,* and *Nitrosyl complexes* are used for the NH_3, CO, H, and NO ligands.

FORMULA INDEX

The Formula Index, as well as the Subject Index, is a cumulative index for Volumes 21–22. The Index is organized to allow the most efficient location of specific compounds and groups of compounds related by central metal ion or ligand grouping.

The formulas entered in the Formula Index are for the total composition of the entered compound, e.g., F_6NaU for sodium hexafluorouranate(V). The formulas consist solely of atomic symbols (abbreviations for atomic groupings are not used) and are arranged in alphabetical order, with carbon and hydrogen always given last, e.g., $Br_3CoN_4C_4H_{16}$. To enhance the utility of the Formula Index, all formulas are permuted on the symbols for all metal atoms, e.g., $FeO_{13}Ru_3C_{13}H_{13}$ is also listed at $Ru_3FeO_{13}C_{13}H_{13}$. Ligand groupings are also listed separately in the same order, e.g., $N_2C_2H_8$, 1,2-Ethanediamine, cobalt complexes. Thus individual compounds are found at their total formula in the alphabetical listing; compounds of any metal may be scanned at the alphabetical position of the metal symbol; and compounds of a specific ligand are listed at the formula of the ligand, e.g., NC for Cyano complexes.

Water of hydration, when so identified, is not added into the formulas of the reported compounds, e.g., $Cl_{0.30}N_4PtRb_2C_4 \cdot 3H_2O$.

$As_2C_{10}H_{16}$, Arsine, o-phenylenebis(dimethyl-, rhodium complex, 21:101

$As_4ClO_2RhC_{21}H_{32}$, Rhodium(1 +), (carbon dioxide)bis[o-phenylenebis(dimethylarsine)]-, chloride, 21:101

$As_4ClRhC_{20}H_{32}$, Rhodium(1 +), bis[o-phenylenebis(dimethylarsine)]-, chloride, 21:101

$As_4Cl_6Nb_2C_{20}H_{32}$, Niobium(III), hexachlorobis[o-phenylenebis(dimethylarsine)]di-, 21:18

$As_6Cl_6Nb_2C_{22}H_{54}$, Niobium(III), hexachlorobis[[2-[(dimethylarsino)methyl]-2-methyl-1,3-propanediyl]bis(dimethylarsine)]-, 21:18

$AuNO_2SC_3H_6$, Gold(I), (L-cysteinato)-, 21:31

$BCuN_6OC_{10}H_{10}$, Copper(I), carbonyl[hydrotris(pyrazolato)borato]-, 21:108

$BCuN_6OC_{16}H_{22}$, Copper(I), carbonyl[tris(3,5-dimethylpyrazolato)hydroborato]-, 21:109

$BCuN_8OC_{13}H_{12}$, Copper(I), carbonyl[tetrakis(pyrazolato)borato]-, 21:110

$BNNaC^2H_3$, Borate(1 −), cyanotri[(^2H)hydro]-, sodium, 21:167

$BN_4RhC_{44}H_{56}$, Rhodium(I), tetrakis(1-isocyanobutane)-, tetraphenylborate(1 −), 21:50

$BN_6C_9H_{10}$, Borate(1 −), hydrotris(pyrazolato)-, copper complex, 21:108

$BN_6C_{15}H_{22}$, Borate(1 −), tris(3,5-dimethylpyrazolato)hydro-, copper complex, 21:109

$BN_8C_{12}H_{12}$, Borate(1 −), tetrakis(pyrazolato)-, copper complex, 21:110

$B_2FeN_6O_6C_{30}H_{34}$, Iron(II), {[tris[μ-[(1,2-cyclohexanedione dioximato)-$O:O'$]diphenyldiborato(2 −)]-$N,N',N'',N''',N'''',$ N'''''}-, 21:112

$B_2N_6O_6C_{30}H_{34}$, Borate(2 −), tris[μ-[(1,2-cyclohexanedione dioximato)-$O:O'$]diphenyldi-, iron complex, 21:112

$BrCoN_4O_3C_5H_6$, Cobalt(III), (carbonato)bis(1,2-ethanediamine)-, bromide, 21:120

$BrCoN_4O_7S_2C_4H_{18} \cdot H_2O$, Cobalt(III), aquabromobis(1,2-ethanediamine)-, dithionate, $trans$-, monohydrate, 21:124

$Br_2N_4PdC_{12}H_{30}$, Palladium(II), [N,N-bis[2-(dimethylamino)ethyl]-N',N'-dimethyl-1,2-ethanediamine]bromo-, bromide, 21:131